NASA SP-242

I0482669

GUIDE TO LUNAR ORBITER PHOTOGRAPHS

Thomas P. Hansen

Langley Research Center

Hampton, Virginia

Prepared by Langley Research Center

Scientific and Technical Information Office   1970
NATIONAL AERONAUTICS AND SPACE ADMINISTRATION
Washington, D.C.

Preface

This document provides information on the location and coverage of each photograph returned by the Lunar Orbiter series of spacecraft. Small-scale maps show the overall coverage of each mission and the areas of common coverage among sites of different missions. Large-scale maps show coverage of the individual photographs at each target area. The characteristics of the cameras and of the various orbital sequences utilized are given for background information pertinent to an understanding of Lunar Orbiter photography.

Introduction

The Lunar Orbiter program initiated in early 1964 consisted of the investigation of the Moon by five identical unmanned spacecraft. Its primary objective was to obtain detailed photographs of the Moon. This document presents information on the location and coverage of all Lunar Orbiter photographs and is one in a series of four NASA Special Publications documenting Lunar Orbiter photography. The others are references 1 to 3. Reference 1 contains 675 photographic plates and provides coverage of the complete Moon with more detail than any other publication. Reference 2 is a collection of approximately 180 selected photographs and portions thereof at enlarged scale, and includes captions for each photograph. Reference 3 shows each named feature on the near side on annotated high-resolution frames from mission IV. It also includes (1) an alphabetical index of features, (2) cross-indexes between listings in the catalog of the University of Arizona and the catalog of the International Astronomical Union which was published in 1935, and (3) listings of named lunar features on the near side covered during missions I, II, III, and V, and their photograph numbers.

The objectives of the Lunar Orbiter program were—

(1) Photography.—To obtain detailed lunar topographic and geologic information of various lunar-terrain types to assess their suitability for use as landing sites by Apollo and Surveyor spacecraft and to increase man's scientific understanding of the Moon.

(2) Selenodesy.—To provide precision trajectory information which would improve the definition of the lunar gravitational field.

(3) Moon environment.—To provide measurements of micrometeoroid and radiation flux in the lunar environment for spacecraft performance analysis.

These objectives were accomplished by the flights of five spacecraft during the 13-month period from August 1966 to September 1967. In addition to references 1 to 3 on Lunar Orbiter photography, the interested reader is directed to references 4 to 7 for results of the program.

The five Lunar Orbiter spacecraft returned over 1654 high-quality photographs taken from lunar orbit. Each spacecraft was similarly equipped with two cameras which operated simultaneously and had the same line of sight but different fields of view and resolutions. The cameras utilized a common supply of 70-mm film and the dual images they recorded are referred to as medium-resolution frames and high-resolution frames.

Of the 1654 Lunar Orbiter photographs, 840 are of areas photographed on the basis of Apollo program requirements and were obtained primarily during missions I, II, and III. They were taken from low flight altitudes and provided detailed coverage of 22 areas located along the equatorial region of the near side of the Moon. The remaining 814 photographs were taken primarily during missions IV and V and include 703 of the near side of the Moon, 105 of the far side of the Moon, and 6 of the Earth. These photographs were taken from flight altitudes ranging from approximately 44 km over the near side to approximately 6000 km over the far side, and provide broad coverage of essentially the entire Moon and detailed coverage at 88 areas on the near side.

This document contains tables and maps which catalog the various types of Lunar Orbiter photography conducted and aid the user in procuring photographs of selected areas. The maps were prepared by the U.S. Air Force Aeronautical Chart and Information Center, in support of preliminary photo analyses performed immediately following each Lunar Orbiter mission.

The National Space Science Data Center (NSSDC) at Goddard Space Flight Center, Greenbelt, Md., is responsible for dissemination of Lunar Orbiter photographs and other scientific data. Scientists requiring high-quality Lunar Orbiter photographs for study can obtain them from that Center. Persons interested in Lunar Orbiter photographs for other reasons should direct their requests to NASA, Public Information Division, Code FP, Washington, D.C. 20546.

Lunar Orbiter Photographic System

A sketch of the photographic system of the spacecraft is shown in figure 1. The system was housed in a pressurized, thermally controlled container, and included the cameras, film and film handling, film processor, and readout equipment and environmental controls. The system was designed to expose, develop, and read out images for transmission to Earth by the communications system.

The two cameras simultaneously placed two discrete frame exposures on a common supply of 70-mm aerial film. Each camera operated at a fixed aperture of $f/5.6$ with controllable shutter speeds of 0.01, 0.02, or 0.04 second. One of the lenses had a 610-mm focal length; the other, an 80-mm focal length.

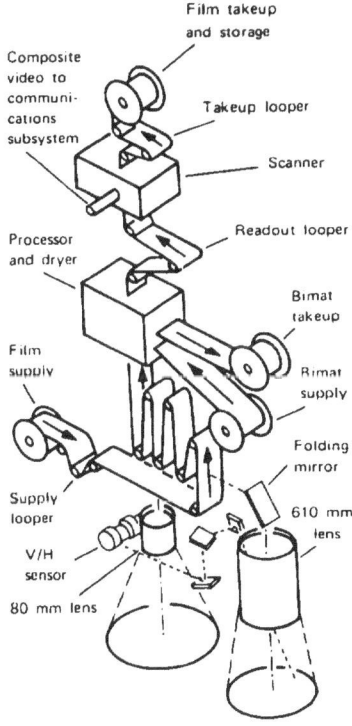

FIGURE 1.—*Photographic subsystem.*

Shutter, platen, and image-motion compensation were provided for each camera; the film, film advance, and shutter operation were common to both. The film was developed onboard by using a method which passed the film into contact with a web that contained a single-solution processing chemical. After the film was dried, it was stored ready to be read out and transmitted to Earth.

Figure 2 shows a schematic of the readout system which used a line-scan tube as the light source for scanning the negative image on the spacecraft film. The line-scan tube electronically scanned the beam of light a distance of 2.667 mm in the lengthwise direction of the film. The sweep of this line across the film was accomplished by a mechanical drive of the scanner lens which focused the line. One traverse of the scanner lens across the film required approximately 22 seconds; during this time the line scan was repeated over 17 000 times. The sections of film that were read out with this type of scan were referred to as framelets and were 2.54 mm wide and over 55 mm long. At this rate, 10 minutes were required to read out one medium-resolution frame, and 34 minutes for one high-resolution frame. The transmitted light was sensed by a photomultiplier tube and the resulting electrical signal was mixed with synchronization and blanking pulses and fed to the communication system modulator for transmission to Earth. The video signal received on Earth was fed into the ground reconstruction electronics (GRE) where it was converted into an intensity-modulated line on the face of a cathode-ray tube. This line was used to expose 35-mm film in a continuous-motion camera to reconstruct the framelets. The scale of the reconstructed framelets (GRE scale) was 7.18 times spacecraft scale; the framelets were approximately 18 mm wide and 40 cm long. The framelets were then reassembled. Medium-resolution frames were reassembled in their entirety; high-resolution frames were reassembled into three component sections.

The video signal was also recorded on magnetic tapes which were subsequently used to make additional 35-mm framelets. These framelets had generally improved tonal qualities over the framelets reconstructed during the missions and were used to make master negatives for use by the NSSDC in providing copies to the public.

The film supply of the spacecraft consisted of 79 meters of unperforated 70-mm Kodak aerial film, type SO-243. This film

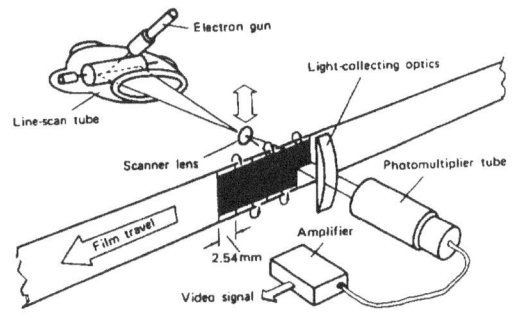

FIGURE 2.—Readout scanner.

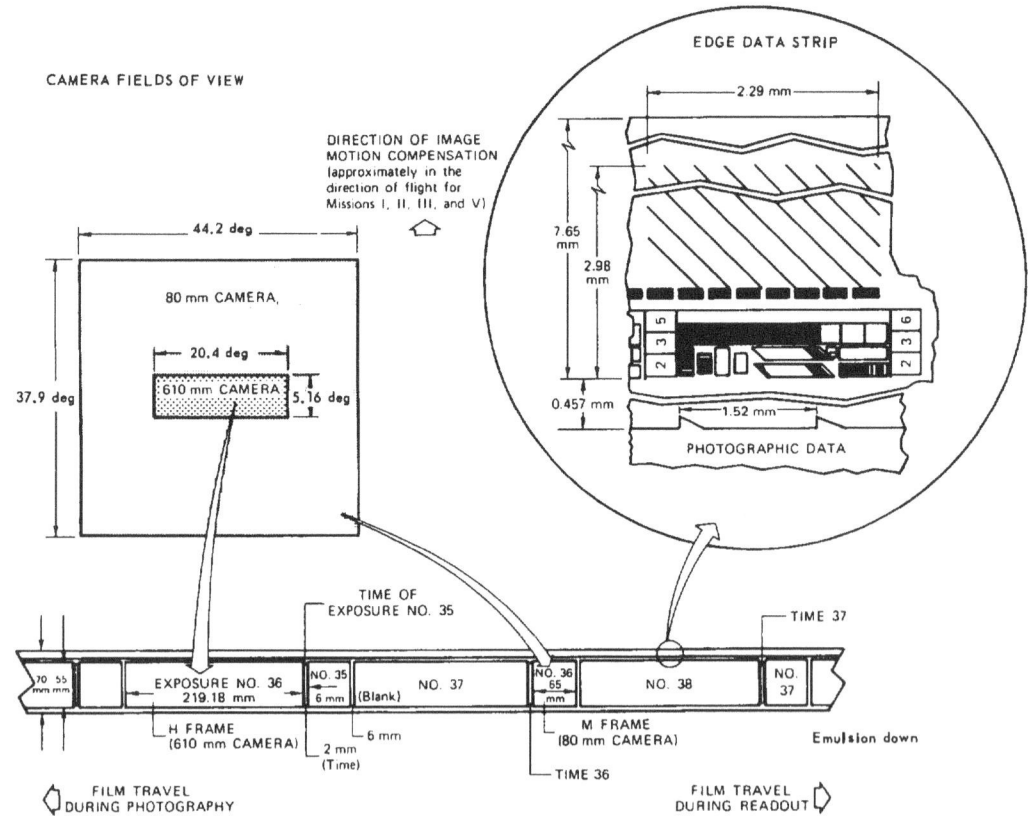

FIGURE 3.—Spacecraft film format.

is a fine-grained, low-speed film with an aerial index of 3.0, which makes it relatively insensitive to space environment radiation. The film was provided with image-motion compensation (IMC) by a velocity/height (V/H) sensor which utilized the 610-mm lens. The V/H sensor also controlled the spacing of shutter operations during multiple-exposure sequences.

The full fields of view (shown in fig. 3) for the 80-mm camera and the 610-mm camera were 44.2° by 37.9° and 20.4° by 5.16°, respectively. The placement of the images of the two cameras on the spacecraft film is shown in figure 3. Images recorded by the 80-mm camera are referred to as medium-resolution frames (M frames); those recorded by the 610-mm camera are referred to as high-resolution frames (H frames).* A folding mirror was employed in the optical path of the 610-mm camera and therefore the H-frame images are reversed left to right with respect to the M frames. Exposure times were recorded on the film by a binary-coded arrangement of lamps. These timing lights were located on the 80-mm camera platen and recorded exposure times to tenths of a second.

The angular resolution of the 610-mm cameras was 4.4 seconds of arc; for the 80-mm cameras, 34 seconds of arc. The resolution of the images recorded by both cameras was 76 lines/mm (spacecraft scale) which translates to an image resolution of approximately 11 lines/mm for the reconstructed 35-mm framelets (GRE scale). The ground resolution of vertical photographs taken from an altitude of 46 km is approximately 1 meter for high-resolution frames and 8 meters for medium-resolution frames.

Prior to launch, the spacecraft film was preexposed along one edge as shown in figure 3. The preexposures included calibration data which were used to monitor inflight system operation and to evaluate final data quality. The preexposed data array included a 0.3 background density to provide a reference level for setting readout gain, diagonal focus lines to indicate optimum readout scanning-spot focus, resolution charts to evaluate readout quality independent of camera image quality, a gray scale for sensitometric calibration, and an identification number. In addition, the film used on missions II, III, IV, and V was preexposed with a geometrical pattern extending across the entire format for geometric calibration purposes.

Photographic Mission Parameters

Table 1 (p. 7) gives the flight log of the five Lunar Orbiter photographic missions and table 2 summarizes the photographic accomplishments. The orbits for each mission were ellipses with orbital parameters selected according to the various tasks of each mission. Mission-site location and the type and extent of required coverage were major considerations in determining the most suitable orbital parameters. Apollo landing sites were to be located within the equatorial region on the near side of the Moon and therefore the first three missions utilized close-in orbits inclined slightly to the lunar equator to provide optimum coverage of these areas. Perilune altitudes for these missions were as low as 44 km, limited primarily by uncertainties in execution errors in maneuvers, uncertainties in the lunar gravitational field and elevations, and the operating range of the V/H sensor. Missions IV and V were devoted to increasing scientific understanding of the Moon and utilized highly elliptical, near-polar orbits for access to areas at high latitudes with proper illumination. Each spacecraft orbited the Moon in the same sense as the rotation of the Moon.

---
*Other terms used in literature in referring to Lunar Orbiter photographs are for medium-resolution frame: wide-angle frame, low-resolution frame, moderate-resolution frame; for high-resolution frame: telephoto frame. This paper uses the words "photograph" and "frame" interchangeably.

During each mission, photography of the near side was conducted near perilune with morning illumination and photography of the far side near apolune with evening illumination. Photographs were sequenced by using one of two exposure interval rates. Spacecraft photographic maneuvers were based on the requirements that the camera axes must point directly at the target position at the midpoint of a sequence and that image-motion compensation is provided (when required) by proper orientation of the spacecraft with respect to the flight-path. These maneuvers usually consisted of a three-axis rotation from the normal Sun-Canopus celestial reference several minutes prior to the picture taking. Each photograph in a sequence would be taken with the camera axes in the same reference after which the spacecraft would be returned to the celestial reference. Since photographs could be taken much faster than they could be processed, a looper having a capacity of 20 dual frames acted as a buffer between the cameras and the readout section. Site photography proceeded from east to west as the Moon rotated under the stationary, inertially fixed orbit of the spacecraft.

Photographic Coverage

AREAS OF PARTICULAR INTEREST TO APOLLO PROGRAM

Photography during Lunar Orbiter missions I, II, and III was conducted primarily to locate and confirm suitable manned landing sites for the Apollo program. The requirements for these sites were as follows:—

Zone of interest.—The sites had to be located within the zone specified by the Apollo program; ±45° longitude and ±5° latitude.

Site locations.—Multiple sites which permitted at least three launch opportunities within any Apollo launch window had to be located. Launch opportunities were anticipated to occur on alternate days; thus, suitably lighted sites separated in longitude by 23°±3° were required. In addition, the capability to launch during each month of the year required sites to be located along both the northern and southern portions of the zone mentioned.

Site characteristics.—Apollo landing sites had to cover an elliptical area with major and minor axes approximately 8 km and 5 km, respectively, and had to be relatively free of protuberances, depressions, or slopes that would constitute a hazard to the Apollo landing vehicles. The landing-approach terrain was to be reasonably unmodulated to accommodate the guidance system of the vehicle.

Areas were originally selected from Earth-based observations that appeared to offer candidate Apollo landing sites. Areas photographed as candidate landing sites were designated primary (P) sites. Prescheduling of P sites left certain periods when photographs had to be taken in order to satisfy film-handling constraints. Sites photographed in compliance with this constraint were designated secondary (S) sites; it should be noted, however, that this designation had relevance only with respect to the mission objectives and the mission plan and not to the value of the photography.

Mission I photography was conducted from three different orbits characterized by the parameters listed in table 1. Nine primary sites concentrated in the southern part of the Apollo zone were photographed. The 610-mm camera failed to operate satisfactorily at close range and consequently most of the high-resolution frames were smeared. The medium-resolution frames were of good quality and provided coverage of extensive areas with an increase in resolution of two orders of magnitude over astronomical photographs. Secondary-site photography provided coverage of numerous areas along the equatorial region on both the near and far sides. The 610-mm camera generally operated satisfactorily during photography of the far side; consequently, both the medium-resolution and high-resolution photographs of these areas were of excellent quality. Ground resolution of the high-resolution frames was approximately 30 meters. Two oblique exposures (nos. 102 and 117) were taken during mission I. Both are very similar in nature, cover approximately the same area, and show views of the crescent Earth and part of the far side of the Moon just beyond the eastern limb (as seen from Earth). In each case both the medium-resolution frame and the high-resolution frame are of good quality.

Mission II photography was conducted from a single orbit having the parameters listed in table 1. Photographic targets were concentrated in the northern part of the Apollo zone of interest and included 13 primary sites and 17 secondary sites. Most of the primary sites were photographed by taking multiple-exposure sequences during consecutive passes. The secondary sites provided equatorial coverage of areas near the equator on both the near and far sides. With the exception of a few photographs which were incompletely read out, no problems were encountered with the operation of the photographic system.

During missions I and II, all the primary sites were photographed by using standard techniques; that is, coverage was obtained by taking sequences of 4, 8, or 16 vertical photographs during one or more passes of the spacecraft over the site. This procedure provided stereoscopic medium-resolution coverage and high-resolution coverage useful primarily for interpretation and photometric analysis. An experiment conducted during mission II determined that even better stereoscopic coverage could be obtained with the 610-mm camera by photographing the same area during two consecutive passes with the camera axes tilted during one of the passes; this type of photography is referred to as convergent photography. The success of this experiment contributed to the decision that mission III would be a site-certification mission. It provided additional coverage of the most promising candidate Apollo landing sites photographed during missions I and II. Other factors which compelled this decision were the need for makeup high-resolution coverage of areas inadequately covered during mission I and the desire to obtain oblique views of the landing sites to simulate the views which would exist during a manned descent to the surface. The 12 primary sites photographed during mission III included 5 areas previously photographed during mission I, 5 areas previously photographed during mission II, and 2 proposed Surveyor landing sites which had been selected on the basis of Earth-based photography. To photograph all these areas under favorable lighting conditions, the inclination of the orbit to the lunar equator was increased from the 12° value used for missions I and II to a value of 21°. Whereas missions I and II employed standard photographic techniques, diverse techniques were used during mission III to take the required vertical, oblique, and converging coverage. Unfortunately, the spacecraft developed trouble with its film advance motor late in the mission and was unable to read out a substantial number of photographs. This mission provided the Apollo program with sufficient information, however, to allow the remaining two missions to concentrate on more expanded scientific objectives.

In summary, 22 areas were photographed during the first three missions in search of Apollo landing sites. On the basis of this photography and data obtained from Surveyor I, eight candidate Apollo landing sites were selected. Although all three types of required coverage had been obtained at only three of these sites, additional coverage obtained later during mission V enabled complete certification of all sites. Table 3 summarizes all photography during missions I, II, III, and V taken in search of Apollo landing sites. The most promising areas were those photographed with all three types of photography indicated. Entries for "Area of interest" begin with the easternmost site and progress westward.

AREAS OF GENERAL INTEREST

Most of the photographs taken during missions I, II, and III were of areas which by nature of their potential use as manned landing sites were smooth and featureless. Although the secondary-site photography of these missions included a substantial number of areas interesting from the standpoint of geology and resulted in some very spectacular views, this photography was scheduled "around" the primary-site photography and was limited to areas located near the equator. The converse was true for missions IV and V, whose primary objective was to increase man's scientific understanding of the Moon.

Mission IV was assigned the task of performing a broad systematic survey of lunar-surface features in order to increase the scientific knowledge of their nature, origin, and processes, and to serve as a basis for selecting sites for more detailed scientific study by subsequent orbital and landing missions. Photography was planned on the basis of the coverage to be obtained by the 610-mm camera. It was desired to obtain vertical high-resolution photographs which would provide monoscopic coverage of the entire near side with a minimum of overlap. This coverage was obtained by taking 5 single-frame sequences on each of 29 consecutive passes. The orbit was highly inclined to the equator (85°) and had a perilune altitude, at the equator, of approximately 2700 km. The spacecraft was oriented with the long dimension of the frames in a north-south direction. Pole-to-pole coverage was obtained by taking, on each pass, four vertical photographs symmetrically spaced about the equator for coverage of the equatorial and temperate regions, and a fifth photograph for coverage of the polar regions. The fifth photograph was used alternately from pass to pass for coverage of the south- and north-polar regions. It was taken slightly off vertical for lighting considerations.

The near-side photography covered the equatorial regions with ground resolutions of approximately 60 meters and the polar regions with ground resolutions of approximately 100 meters. The field of the 80-mm camera encompassed nearly the entire lunar disk. The ground resolution of the medium-resolution frames is comparable to the best obtainable from astronomical photography—on the order of ½ kilometer.

Many photographs, taken early in the mission, were severely degraded during a period when a thermal door to the cameras failed to operate properly. In some cases the door failed to open and therefore the expected photographs were unexposed. In other cases, the photographs were degraded because of condensation on the camera windows. All the areas covered by these degraded photographs were rephotographed toward the end of the mission by six sequences taken near apolune. As in the case of all photographs taken near apolune, these photographs were taken with evening illumination.

Mission IV coverage of the far side was obtained by five sequences taken near apolune and by a number of the near-perilune sequences. The photographs taken near apolune consisted of seven medium-resolution frames (two of which were severely degraded); the high-resolution frames covered essentially unilluminated areas. Medium-resolution frames taken near perilune (with morning illumination) provided the more significant far-side coverage during mission IV. The photographs taken on the first pass covered extensive areas beyond +90° longitude, and each medium-resolution frame taken on the polar sequences, although centered on the near side, provided coverage which extended beyond the polar caps and on to the far side.

The primary objective of mission V was to photograph 36 areas of particular scientific interest on the near side. Photography was also required to complete the Apollo requirements and to complete the far-side coverage. This combination of requirements necessitated two orbital changes. The orbital parameters are given in table 1. Photographic altitudes for the near side were on the order of 100 km to 250 km. These altitudes, which were two to five times greater than those used for the near-side photography during missions I, II, and III, were required in order to provide adequate areal coverage and acceptable ground resolution of each of the numerous sites with the limited film supply of the spacecraft. In addition, most of the remaining Apollo requirements were for converging coverage. Thus, the increase in altitude was desirable, since it enabled these photographs to be taken with less cross-track tilt than had been utilized previously.

Mission V was executed precisely as planned and accomplished each of its assigned objectives. One dual frame was also taken which shows a view of a nearly full Earth.

In summary, photography for purposes other than locating or confirming Apollo landing sites was taken during each of the five missions. It consists of low-altitude photography of the near side taken during missions I, II, III, and V; and high-altitude photography taken during each of the missions. The low-altitude photography provided detailed coverage of 88 areas from altitudes ranging from approximately 44 to 250 km; this photography is summarized in table 4. The photography at a selected number of these areas is summarized in table 5; features photographed are in alphabetical order. The high-altitude photography provided broad coverage of essentially the entire Moon from altitudes ranging from approximately 1350 to 6000 km. Whereas mission IV alone provided the broad coverage of the near side, each mission contributed to the broad coverage of the far side.

MAPS

Figures 4 to 11 are small-scale maps showing all the areas photographed during each mission, with the exception of the areas covered by the medium-resolution frames from mission IV. Figure 4 (p. 22) is a composite plot for missions I, II, III, and V and indicates the missions during which any given area was photographed. Figures 5 to 9 break down the coverage shown in figure 4 and present separately the coverage obtained during each mission. The photographic coverage obtained during mission IV is shown in figures 10 and 11.

Figures 4 to 11 show, where the scale permits, the areas covered by individual photographs. Where scale limitations precluded showing these areas, they show only the envelope of the total coverage at each site. For each of these sites, the areas covered by the individual photographs are shown in figures 12 to 15, which are large-scale maps.

Thus, figures 4 to 15 permit one to determine all photographs covering a given area. One should first consider the coverage of missions I, II, III, and V and, secondly, that of mission IV.

Missions I, II, III, and V.—Figure 4 shows the total area photographed during each of these missions. It should be noted that this figure presents only the envelope of the total coverage by a given mission in any region. Where an area of near-vertical or converging coverage is contained within an area of oblique coverage photographed during the same mission, only the boundary of the oblique coverage is indicated. Thus, the boundaries of coverage for sites IIP-8, IIIP-7, IIIP-8, IIIP-10, IIIP-11, IIIP-12, IIIS-15, IIIS-16, V-8, V-11, V-12, V-16, and V-18 cannot be separately identified in figure 4. However, they are individually outlined in figures 5 to 9, which show the area covered at each site with the Lunar Orbiter site designa-

tion. Figures 5, 6, and 7 pertain to missions I, II, and III, respectively; figures 8 and 9 pertain to mission V. The coverage shown is the envelope of coverage of the medium-resolution frames for the near-side sites, and with the exception of mission I sites IS-3 and IS-9, the coverage of individual photographs for the far-side sites.

Table 6 gives the exposures allocated to each site for these missions; table 7 is a permuted form of table 6, and indicates the site to which each exposure was assigned. Table 8 lists the mission I, II, III, and V sites for which photographs were incompletely read out or degraded.

Figures 12, 13, 14, and 15 are photographic indexes of all near-side sites, except site IS-1, for missions I, II, III, and V, respectively. They show individual photographic outlines portrayed on the U.S. Air Force Aeronautical Chart and Information Center (ACIC) series of lunar charts (Lunar Aeronautical Charts (LAC) or Apollo Intermediate Charts (AIC)). The photographic outlines are accompanied by numbers which uniquely identify the photographs and which should be used in ordering photographs from NSSDC. The photographic outlines were determined by ACIC personnel who matched the photographic images to the shaded relief features on the charts. Thus, the inferred coordinates of the corners of the photographs and the features contained therein are only as accurate as the charts.

At many sites, especially the sites photographed for Apollo, the high-resolution frames have not been indexed. They were not indexed because there was insufficient detail on the base maps with which to make an image match, the photographs were either not read out or were degraded, or their inclusion would have cluttered the figure. The approximate coverage of these frames can be determined, for vertical or near-vertical photography, by scaling the fields of the two cameras, shown in figure 3, to the map scale of the photographic index. To determine which photographs were degraded or incompletely read out, reference should be made to table 8.

At multiple-exposure sites, the exposure numbers increase from west to east for missions I, II, and III sites (figs. 12, 13, 14), and from south to north for mission V sites (fig. 15). The maps are oriented in the standard aeronautical convention with north at the top. They incorporate the selenographic coordinate system with east (positive) and west (negative) longitudes measured from the central meridian at Sinus Medii and the longitudes increase in magnitude to 180 at the center of the far side.

Mission IV.—Whereas missions I, II, III, and V were assigned to photograph selected areas, mission IV was assigned to photograph broad areas and to cover the entire near side. Both medium- and high-resolution frames from mission IV cover the entire near side, and the medium-resolution frames provide the only coverage of some regions of the far side. The maps of coverage of these photographs are presented independently of those from the other missions.

With the exception of two small areas near the poles, any area which figure 4 indicates was not photographed during missions I, II, III, and V was photographed during mission IV. Figure 10 shows the area covered by each mission IV high-resolution frame. Any area for which neither figure 4 nor figure 10 indicates as having been photographed was covered only by mission IV medium-resolution frames. Figure 11 shows the area covered by a selected number of these photographs (or portions thereof). In most cases the near-side areas covered by these photographs are not shown. The outlines shown in figures 10 and 11 are accompanied by the appropriate exposure number.

Table 9 gives the selenographic distribution of mission IV exposures. Table 10 is a permuted form of table 9 and indicates the site to which each exposure was assigned. Table 11 summarizes all mission IV photographs incompletely read out or degraded.

Map summary.—For any given area, the photographs covering that area are determined as follows:

Refer to figure 4 to determine whether the area was photographed during missions I, II, III, and V and, if so, during which missions(s). Then, depending on the mission(s), refer to the appropriate figure(s) among figures 5 to 9 to determine the site(s). If the area in question is on the near side, the site number is used to locate the photographic index for that site in figures 12 to 15. If the area in question is on the far side, refer to table 6 to determine the exposure number(s).

Refer to figures 10 and 11 to determine whether the area was photographed during mission IV and, if so, by which photograph(s). These figures show the area covered by individual photographs (and the exposure number) for all the high-resolution frames, but only for a selected number of medium-resolution frames. The approximate locations of the principal ground point and condition of the photograph, for the medium-resolution frames not considered in figures 10 and 11, are given in tables 9 and 11, respectively.

Copies of Photographs

Each Lunar Orbiter spacecraft was supplied with sufficient film to record as many as 426 photographs—213 pairs of medium-resolution and high-resolution frames. The negative images on the spacecraft film were read out in parts, termed "framelets," and reconstructed on Earth on 35-mm film as positive images of the Moon at a scale (GRE scale) 7.18 × spacecraft scale. The framelets were then used to make reassembled frames in various forms.

20- BY 24-INCH SECTIONS

The framelets reconstructed in the GRE represent the original flight data and are designated as zero-generation positives. The original framelets (or copies) were reassembled and contact printed on to 20- by 24-inch sheet film. One medium-resolution frame required just one 20- by 24-inch section, whereas the high-resolution frame required three component sections. All Lunar Orbiter photographs have been reassembled into a 20- by 24-inch format, with the exception of the smeared high-resolution frames of mission I.

By using second-generation duplicate positives of the original flight data, the U.S. Army Topographic Command (TOPOCOM) prepared third-generation 20- by 24-inch master negatives for all photographs from missions III, IV, and V and for the high-resolution frames from mission II. These negatives were made to provide quick copies for Government agencies for interpretation and mission planning and to provide the National Space Science Data Center (NSSDC) with material from which early copies could be made generally available to the scientific community.

At the completion of the Lunar Orbiter program, the NASA Langley Research Center (LRC) produced an improved set of 20- by 24-inch negatives from which high-quality copies could be made and disseminated by the NSSDC. The video tapes were used to generate a new set of positive framelets which had generally improved tonal qualities over those secured during the missions. These positive framelets were made by electronic preprocessing of the video signal prior to input to the GRE. (However, because the video signal was intentionally distorted prior to input to the GRE, the 35-mm film exhibits density variations which are not accurate representations of the true lunar reflectance properties and should not, therefore, be used for densitometric or photometric analysis.) The positive framelets thus obtained were reassembled and contact printed on to 20- by 24-inch sheet film to make first-generation master negatives. This procedure was followed for all photographs except those not graded A or B in tables 8 and 11. Each 20- by 24-inch section is labeled with a photo number consisting of mission number, a Roman numeral; exposure number, an Arabic numeral; and frame type, M (medium-resolution frame) or H (high-resolution frame). Sections of high-resolution frames are additionally labeled with subscripts 1, 2, or 3 following the photo number to distinguish the component sections. For example, the sections labeled V–141M and V–141H$_2$ are mission V medium-resolution frame no. 141 and the center section of mission V high-resolution frame no. 141, respectively. For the photographs listed in table 12, the video tapes were replayed additional times to produce 35-mm film with optimum detail in the highlight areas or the lowlight areas. The photographs made from reassemblies of this film are additionally labeled with "SP," indicating a special play for highlight areas, or "SP–1," indicating a special play for lowlight areas.

Table 13 gives some characteristics of Lunar Orbiter vertical photographs. Values given for the photographic scale apply to the 35-mm framelets reconstructed in the GRE and also to the 20- by 24-inch sections. The ground resolutions given are in direct proportion to the altitudes given. The reassembly code given for the high-resolution frames is useful for orienting the photographs. The long axis of all photographs is oriented either in a primarily north-south or an east-west direction. With the edge data at the top, the left, center, and right sections (of a high-resolution frame for those frames reassembled at the Langley Research Center (LRC)) are numbered 1, 2, and 3, respectively. The reassembly code tells which of these sections provides the northernmost (N) or easternmost (E) coverage. It applies only for frames reassembled at LRC. (TOPOCOM numbered the three-component sections of a high-resolution frame in the reverse order: sections 1, 2, and 3 in the LRC convention are sections 3, 2, and 1, respectively, in the TOPOCOM convention.)

SOURCE OF COPIES

The results of all space science flight experiments are made available through the National Space Science Data Center (NSSDC). Copies of all Lunar Orbiter photographs and background information including photographic system calibrations and photographic supporting data are available from the NSSDC. For further information, scientists located within the United States should address their inquiries to—

> National Space Science Data Center
> Code 601.4
> Goddard Space Flight Center
> Greenbelt, Md. 20771

Scientists from abroad, to—

> World Data Center A
> Rockets and Satellites
> Code 601
> Goddard Space Flight Center
> Greenbelt, Md. 20771, U.S.A.

In ordering copies, the photographs should be specified by mission number, exposure number, and frame type (M or H). When interested in a particular section of a high-resolution frame, the position of that section relative to the central section—northern, eastern, etc.—should be stated. The quantity of Lunar Orbiter photographs available from the NSSDC, in terms of 20- by 24-inch sections, is given in table 14.

TABLE 1.—*Lunar Orbiter Flight Log*

	Mission I	Mission II	Mission III	Mission IV	Mission V
Launch:					
Date	8/10/66	11/6/66	2/5/67	5/4/67	8/1/67
Hr:min (GMT)	19:26	23:21	01:17	22:25	22:23
Injection into lunar orbit:					
Date	8/14/66	11/10/66	2/8/67	5/8/67	8/5/67
Hr:min (GMT)	15:43	22:58	22:03	15:17	16:49
Photographic dates:					
First exposure	8/18/66	11/18/66	2/15/67	5/11/67	8/6/67
Last exposure	8/29/66	11/25/66	2/23/67	5/25/67	8/18/67
Mission termination:					
Date of impact	10/29/66	10/11/67	10/9/67	*7/17/67	1/31/68
Hr:min (GMT)	13:29	07:17	10:27	06:30	07:58
Impact location:					
Longitude, deg	160.71 E	119.13 E	92.70 W	≈26 W	83.04 W
Latitude, deg	6.35 N	2.96 N	14.32 N		2.79 S
Orbital parameters utilized for photography:					
First set:					
Perilune altitude, km	189	50	55	2706	195
Apolune altitude, km	1866	1853	1847	6114	6028
Inclination, deg	12.16	11.89	20.91	85:48	85.01
Period (hr:min)	3:37	3:28	3:28	12.01	8:27
Exposures taken	5 to 42	5 to 215	5 to 215	5 to 196	5 to 22
Second set:					
Perilune altitude, km	56				100
Apolune altitude, km	1853				6067
Inclination, deg	12.05		Not applicable		85.61
Period (hr:min)	3.29				8:21
Exposures taken	44 to 133				24 to 30
Third set:					
Perilune altitude, km	40				99
Apolune altitude, km	1817				1500
Inclination, deg	12.00		Not applicable		84.76
Period (hr:min)	3:26				3:12
Exposures taken	134 to 215				31 to 217

*Last communication with spacecraft. Date of impact estimated at 10/31/67.

TABLE 2.—*Number of Photographs Obtained*

Mission	Number of sites	Number of exposures	Number of photographs obtained —					
			Medium-resolution frames			High-resolution frames		
			Useful for interpretation		Not useful for interpretation or not read out	Useful for interpretation		Not useful for interpretation or not read out
			Complete frames	Partial frames		Complete frames	Partial frames	
Apollo								
Mission I	9	136	136	0	0	0	0	136
Mission II	13	184	180	1	3	175	7	2
Mission III	18	162	116	4	42	102	31	29
Mission V	9	44	44	0	0	44	0	0
Subtotal	49	526	476	5	45	321	38	167
General interest; near side								
Mission I	18	57	57	0	0	6	0	51
Mission II	13	23	23	0	0	23	0	0
Mission III	23	47	35	1	11	35	3	9
Mission V	36	130	129	0	1	129	1	0
Mission IV (subtotal)	148	165	117	3	45	132	9	24
Missions I, II, III V (subtotal)	90	257	244	1	12	193	4	60
General interest; far side								
Mission I	2	11	11	0	0	6	0	5
Mission II	4	4	4	0	0	4	0	0
Mission III	2	2	1	0	1	1	1	0
Mission IV	6	7	5	0	2	0	0	7
Mission V	23	37	37	0	0	35	0	2
Subtotal	37	61	58	0	3	46	1	14
Earth								
Earth	-----------	3	3	0	0	3	0	0
Grand total	324	1012	898	9	105	695	52	265

TABLE 3.—*Photography in Search of Apollo Landing Sites*

(a) Photographic information

Area of interest			Vertical and near-vertical photography				Converging photography			Oblique photography (west looking)	
Vicinity of search site*	Approximate location		Site	First sequence, exposures	Second sequence, exposures	Note reference in table 3(b)	Site	Near-vertical sequence, exposures	Convergent sequence, exposures	Site	Exposure
	Longitude	Latitude									
IP-1	42° E	1° S	IP-1	52 to 67	-----------	2	IIIP-2	35 to 32	33 to 36	V-3.1	38 (zero-phase photo)
			IIIP-2	25 to 32	-----------	4	V-8	44 to 47	48 to 51		
			V-8	44 to 47	-----------	6				V-6	42
IIP-1	37° E	4° N	IIP-1	5 to 20	-----------	2					
IP-2	36° E	0°	IP-2	68 to 83	-----------	2					
IIP-2	34° E	2° N	IIP-2	35 to 42	-----------	2	V-11	55 to 58	59 to 62	V-9.1	52
			IIIP-1	5 to 20	-----------	4					
IP-3	26° E	1° N	IP-3	85 to 100	-----------	2	IIIP-5	60 to 67	52 to 59	V-13	64
			IIP-6	76 to 83	84 to 91	2	V-16	71 to 74	75 to 78		
			IIIP-4	44 to 51	-----------	4					
			IIIP-5	60 to 67	-----------	4					
IIP-5	25° E	3° N	IIP-5	67 to 74	-----------	2					
IIP-3	21° E	4° N	IIP-3	43 to 50	51 to 58	2					
IIIP-3	21° E	2° N	IIIP-3	40 to 43	-----------	5					
IIIP-6	21° E	0°	IIIP-6	68 to 71	-----------	5					
IIP-4	16° E	4° N	IIP-4	59 to 66	-----------	2					
IP-4	14° E	0°	IP-4	105 to 112	-----------	3					
IP-5	1° W	0°	IP-5	118 to 133	-----------	2	IIIP-7	94 to 101	86 to 93	IIIS-11	84
			IIP-8	113 to 120	121 to 128 (129 to 136, 3d sequence)	2	V-27	108 to 111	112 to 115		
			IIIP-7	94 to 101	-----------	4					
IP-6	2° W	4° S	IP-6	141 to 148	-----------	3					
IIP-7	2° W	2° N	IIP-7	96 to 103	104 to 111	2					
IIP-9	13° W	1° N	IIP-9	138 to 145	-----------	2					
IIP-11	20° W	0°	IIP-11	163 to 170	171 to 173	2				IIIS-21	120
			IIIP-8	124 to 131	-----------	4					
			IP-7	157 to 172	-----------	2	IIIP-9	145 to 152	137 to 144	IIIS-24	136
IP-7	22° W	3° S	IIIP-9	145 to 152	153 to 160	4					
			IIP-10	146 to 153	154 to 161	2					
IIP-10	27° W	3° N	IIP-12	179 to 186	187 to 194	2					
IIP-12	34° W	2° N	IP-8.1	176 to 183	-----------	2	V-42	169 to 172	173 to 176	IIIS-27	171
IP-8.1	36° W	3° S	IIIP-11	173 to 180	-----------	4					
IIP-13	41° W	2° N	IIP-13	197 to 204	205 to 212	2		205 to 212 (IIP-13)	163 to 170 (IIIP-10)	IIIS-25	161
IP-9.2	43° W	2° S	IP-9.2	184 to 199	200 to 215	2	IIIP-12	185 to 192	205 to 212	IIIS-28	172
			IIIP-12	181 to 184	185 to 200 (201 to 204, 3d sequence)						

*Search site: Area of vertical coverage photographed in search of Apollo landing sites. Candidate Apollo landing sites selected on the basis of this photography were certified by the additional vertical, converging, and oblique photography listed. Search sites which did not reveal areas suitable for Apollo were not rephotographed except in an incidental manner. These areas of common coverage are not indicated in this table but may, however, be determined by reference to the index maps.

(b) Vertical photography

Note reference	Exposure internal rate	Stereoscopic coverage, medium-resolution frames		Monoscopic coverage, high resolution frames	
		Forward overlap, percent	Lateral overlap,* percent	Forward overlap, percent	Lateral overlap,* percent
2	Fast	88	66	Continuous	11
3	Slow	52	66	Discontinuous	11
4	Fast	88	42	Continuous	Discontinuous
5	Slow	52	42	Discontinuous	Discontinuous
6	Fast	88	Not applicable	Continuous	Not applicable

*Lateral overlap given for photographs taken on adjacent sequences.

TABLE 4.—*Photography of Areas of General Interest—Near Side*

Site	Approximate center of coverage		Type of photography [a]	Remarks	Site	Approximate center of coverage		Type of photography [a]	Remarks
	Longitude	Latitude				Longitude	Latitude		
IS-1b	90° E	1° S	NV, 16f, 4f	Mare Smythii	V-29	3° W	12° N	NV, 4f	Rima Bode II
IS-2b	72° E	2° N	NV		IS-12b	5° W	3° N	NV	
V-1	61° E	26° S	NV, 4f	Petavius	IIIS-15	6° W	0°	NV, 4s	Near Schröter, north of Mösting
IS-4b	60° E	1° N	NV		IIIS-16	6° W	0°	NV	Mösting
V-2.1	58° E	20° S	NV	Petavius B	IIIS-18	8° W	3° S	NV, 4s	Mösting C
V-4	53° E	32° S	NV	Stevinus A	IIIS-14	8° W	6° N	Oblique	Candidate Surveyor site
IS-5b	50° E	2° N	NV,*	Taruntius	V-30	11° W	42° S	NV, 4f	Tycho
IIIS-1b	47° E	1° S	NV, 4f	Messier and Messier A	IIS-10.2	11° W	13° N	NV	Gambart C, thermal anomaly
V-5.1	42° E	2° S	Oblique	Messier	V-32	11° W	13° N	NV, 4s	Erathosthenes
IIS-1	42° E	3° N	NV, 4f		IS-13b	15° W	2° N	NV	Gambart
IS-6b	40° E	3° N	NV*		V-33	15° W	6° N	NV	Area of Copernicus CD
IIS-2	37° E	3° N	---------	Experiment on convergent photography	V-34	16° W	8° S	NV, 4f	Fra Mauro
V-12	34° E	1° S	NV	Censorinus	V-35	16° W	14° N	NV, 4s	Copernicus secondaries
IS-7b	32° E	5° N	NV*	I-48M shows domes near Maskelyne A	IIIS-23	17° W	4° S	NV, 4s	Fra Mauro
					IS-15b	17° W	0°	NV	
V-14	30° E	22° N	NV, 4f	Littrow	V-36	18° W	7° N	NV, 4f	Copernicus H
IIIS-8	27° E	14° S	Oblique	Theophilus	IS-14	20° W	1° N	NV	
V-10	26° E	30° S	Oblique	Altai Scarp	IIS-12	20° W	10° N	Oblique	Copernicus, northerly oblique
V-15.1	26° E	17° N	NV	Dawes	V-37	20° W	10° N	NV, 8f	Copernicus
IIIS-5b	25° E	1° S	Oblique	Moltke	IS-19b	22° W	5° S	NV	
V-18	19° E	2° N	NV, 4f	Dionysius	IIIS-22	22° W	1° N	NV	Candidate Surveyor site
IIIS-9	18° E	2° S	NV	Delambre	V-38	22° W	33° N	NV, 4f	Imbrian flows
IS-8b	17° E	3° N	NV	Dionysius	IS-16b	24° W	0°	NV	Near Reinhold, grooves and chain craters radial to Copernicus
V-19	14° E	15° S	NV	Abulfeda crater chain					
IIIS-10	14° E	2° S	NV, 4s	Candidate Surveyor landing site	IIS-11	27° W	4° N	NV	Southwest of Copernicus near Hortensius
V-21	14° E	39° N	NV, 4f	South of Alexander					
IIS-8	13° E	0°	NV		IIIS-20	27° W	12° N	Oblique	Hortensius domes
IS-10b	9° E	2° N	NV		IS-17b	30° W	1° N	NV	
IIIS-6b	9° E	13° N	Oblique	Hyginus Rilles	V-40	31° W	12° N	NV, 4f	Tobias Mayer dome
V-22	9° E	20° N	NV, 4f	Sulpicius Gallus Rilles	IS-21b	35° W	4° S	NV	
IIIS-7b	7° E	4° N	NV, 4s	Vicinity of Dembowski	IS-18b	36° W	0°	NV	
V-23.2	6° E	8° N	NV, 4f	Hyginus Rilles	IIIS-26	36° W	11° N	Oblique	Kepler
IIS-6	5° E	4° N	NV	Near Triesnecker	V-41	37° W	31° S	NV	Vitello
IIIS-17	4° E	5° S	NV, 4s	Candidate Surveyor site, floor of Hipparchus	V-43.2	40° W	18° S	NV, 4f	Gassendi
					V-45.1	41° W	36° N	NV, 4f	Jura Domes
V-24	4° E	5° S	NV, 4f	Hipparchus	IIS-13	43° W	3° N	NV	Braided ridge southwest of Kepler
V-26.1	4° E	36° N	NV, 4s	Hadley Rille	V-46	43° W	27° N	NV, 8f	Harbinger mountains
IIIS-13	2° E	11° N	Oblique	Murchison and Pallas	V-48	47° W	23° N	NV, 8f	Aristarchus
IIS-9	1° E	2° N	NV	Sinus Medii, southwest of Triesnecker	V-49	49° W	25° N	NV, 4f	Cobra Head
					IS-20b	51° W	4° S	NV	
IIS-7	1° W	3° S	Oblique	Sinus Medii, southerly oblique	IIS-15	52° W	12° N	Oblique	Marius, northerly oblique
V-25	2° W	46° N	Oblique	Alpine valley	V-50	52° W	28° N	NV, 4f	Aristarchus plateau
V-31	2° W	50° N	NV, 4f	Sinuous rille east of Plato	IIS-16	54° W	3° N	NV	South of Reiner
V-28	3° W	14° S	NV, 4f	Alphonsus	IIS-17	55° W	13° N	Oblique	Reiner Gamma
IIIS-19	3° W	4° S	NV, 4s	Candidate Surveyor site, Flammarion	V-51b	56° W	13° N	NY, 8f	Marium Hills
					IIIS-29	62° W	10° S	Oblique	Damoiseau
					IIIS-30	64° W	12° N	Oblique	Cavalerius, Luna 9 area
					IIIS-31	67° W	1° N	NV	Floor of Hevelius

[a] Type of photography:
NV, vertical or near vertical
Oblique

xxf or xxs (applies to multiple exposure sequences). The number of exposures taken per sequence followed by the exposure interval rate; f, fast rate to give 88 percent forward overlap between consecutive medium-resolution frames and continuous high-resolution coverage; and s, slow rate to give 52 percent forward overlap between consecutive medium-resolution frames and discontinuous high-resolution coverage.

[b] Sites at which photographs were incompletely read out or secured in degraded form. See table 8.

* Photographs of these sites were taken on separate orbits having different orbital parameters. Although they were taken independently of each other, they provided continuous coverage of specific areas.

TABLE 5.—*Sites of Selected Areas of Special Interest*

Feature	Site	Feature	Site
Near-vertical photography		Near-vertical photography—Continued	
Abulfeda	V–19	Sinus Medii	IIS–9
Alphonsus	V–28	Messier	IIIS–1
Aristarchus	V–48	Messier A	IIIS–1
Aristarchus Plateau	V–50	Moltke	IIIS–5
Rima Bode II	V–29	Mösting	IIIS–16
Censorinus	V–12	Mösting C	IIIS–18
Cobra Head	V–49	Petavius	V–1
Copernicus	V–37	Petavius B	V–2.1
Copernicus CD	V–33	Mare Smythii	IS–1
Copernicus H	V–36	Stevinus A	V–4
Copernicus Secondaries	V–35	Sulpicius Gallus Rilles	V–22
Dawes	V–15.1	Taruntius	IS–5
Delambre	IIIS–9	Tobias Mayer Dome	V–40
Dionysius	V–18	Tycho	V–30
Dionysius	IS–8		
Eratosthenes	V–32	Oblique photography	
Fra Mauro	V–34		
Fra Mauro	IIIS–23		
Gambart	IS–13	Alpine Valley	V–25
Gambart C	IIS–10.2	Altai Scarp	V–10
Gassendi	V–43.2	Cavalerius	IIIS–30
Hadley Rille	V–26.1	Copernicus	IIS–12
Harbinger Mountains	V–46	Damoiseau	IIIS–29
Hevelius (floor)	IIIS–31	Hortensius Domes	IIIS–20
Hipparchus	V–24	Hyginus Rille	IIIS–6
Hyginus Rille	V–23.2	Kepler	IIIS–26
Imbrian Flows	V–38	Marius Hills	IIS–15
Jura Domes	V–45.1	Sinus Medii	IIS–7
Littrow	V–14	Messier	V–5.1
Marius Hills	V–51	Murchison	IIIS–13
		Pallas	IIIS–13
		Reiner Gamma	IIS–17
		Theophilus	IIIS–8

TABLE 6.—*Exposures Allocated to Each Site for Missions I, II, III, and V*

Mission I		Mission II		Mission III		Mission V	
Site	Exposures	Site	Exposures	Site	Exposures	Site	Exposures
Near side							
IP-1*	52 to 67	IIP-1*	5 to 20	IIIP-1*	5 to 20	V-1	33 to 36
IP-2*	68 to 83	IIP-2	35 to 42	IIIP-2*	25 to 36	V-2.1	37
IP-3*	85 to 100	IIP-3	43 to 58	IIIP-3*	40 to 43	V-3.1	38
IP-4*	105 to 112	IIP-4	59 to 66	IIIP-4*	44 to 51	V-4	40
IP-5*	118 to 133	IIP-5	67 to 74	IIIP-5*	52 to 67	V-5.1	41
IP-6*	141 to 148	IIP-6	76 to 91	IIIP-6*	68 to 71	V-6	42
IP-7*	157 to 172	IIP-7	96 to 111	IIIP-8	124 to 131	V-8	44 to 51
IP-8.1*	176 to 183	IIP-8	113 to 136	IIIP-9	137 to 160	V-9.1	52
IP-9.2*	184 to 215	IIP-9	138 to 145	IIIP-10	163 to 170	V-10	54
IS-1*	5 to 24	IIP-10	146 to 161	IIIP-11	173 to 180	V-11	55 to 62
IS-2*	25, 26, 27	IIP-11*	163 to 178	IIIP-12	181 to 212	V-12	63
IS-4*	29, 33, 34	IIP-12	179 to 194	IIIS-1*	21 to 24	V-13	64
IS-5*	31, 32, 44	IIP-13	197 to 212	IIIS-3*	38	V-14	66 to 69
IS-6*	41, 50, 51	IIS-1	21 to 24	IIIS-4*	39	V-15.1	70
IS-7*	42, 46 to 49	IIS-2	25 to 32	IIIS-5*	72	V-16	71 to 78
IS-8*	84	IIS-6	92	IIIS-6*	73	V-18	80 to 83
IS-10*	103	IIS-7	93	IIIS-7*	74 to 77	V-19	84
IS-12*	113, 114	IIS-8	94	IIIS-8*	78	V-21	86 to 89
IS-13*	134, 135	IIS-9	95	IIIS-9	79	V-22	90 to 93
IS-14*	137, 139, 140	IIS-10.2	112	IIIS-10	80 to 83	V-23.1	94 to 97
IS-15*	138	IIS-11	137	IIIS-11	84	V-24	98 to 101
IS-16*	149, 151	IIS-12	162	IIIS-13	85	V-25	102
IS-17*	150	IIS-13	195	IIIS-14	102	V-26.1	104 to 107
IS-18*	153 to 156	IIS-15	213	IIIS-15	103 to 106	V-27	108 to 115
IS-19*	173	IIS-16	214	IIIS-16	107	V-28	116 to 119
IS-20*	174	IIS-17	215	IIIS-17	108 to 111	V-29	120 to 123
IS-21*	175			IIIS-18	112 to 115	V-30	125 to 128
				IIIS-19	116 to 119	V-31	129 to 132
				IIIS-20	123	V-32	133 to 136
				IIIS-21	120	V-33	137
				IIIS-22	122	V-34	138 to 141
				IIIS-23	132 to 135	V-35	142 to 145
				IIIS-24	136	V-36	146 to 149
				IIIS-25	161	V-37	150 to 157
				IIIS-26	162	V-38	159 to 162
				IIIS-27	171	V-40	164 to 167
				IIIS-28	172	V-41	168
				IIIS-29	213	V-42	169 to 176
				IIIS-30	214	V-43.2	177 to 180
				IIIS-31	215	V-45.1	182 to 185
						V-46	186 to 193
						V-48	194 to 201
						V-49	202 to 205
						V-50	206 to 209
						V-51*	210 to 217
Far side[b]							
IS-3*	28, 30, 35 to 40	IIS-3	33	IIIS-2*	37	VA-1	5 to 12
IS-9*	102[c], 115, 116, 117[c], 136	IIS-4	34	IIIS-21.5	121	VA-2	13 to 20
		IIS-5	75			VA-3	21
		IIS-14	196			VA-4	22
						VA-6	24
						VA-7.1	25
						VA-8	26
						VA-10	28
						VA-11.2	29
						VA-12	30
						VA-13	31
						VA-14	32
						VA-15	39
						VA-16.1	43
						VA-17.1	53
						VA-18.1	65
						VA-19	79
						VA-20	85
						VA-21	103
						VA-22	124
						VA-23	158
						VA-24	163
						VA-25	181

[a] Sites at which photographs were incompletely read out or secured in degraded form. See table 8.

[b] Photographs centered on the far side were all taken obliquely except for the following near-vertical exposures: for mission I, 28, 30, 35, 36, 37, 38, 39, 40, 115, 116, 136; for mission II, 33, 196.

[c] Earth photographs. Lunar Orbiters took 3 exposures of Earth. Mission I exposures 102 and 117 yielded photographs showing a crescent Earth and an oblique view of the far side of the Moon just beyond its eastern limb (as seen from Earth). Mission V, exposure 27 (designated as site VA-9), yielded photographs showing a nearly full Earth.

TABLE 7.—*Assignment of Exposures for Missions I, II, III, and V*

Mission 1		Mission II		Mission III		Mission V	
Site	Exposure number(s)	Site	Exposure number(s)	Site	Exposure number(s)	Site	Exposure number(s)
IS–1	5 to 24	IIP–1	5 to 20	IIIP–1	5 to 20	VA–1	5 to 12
IS–2	25 to 27	IIS–1	21 to 24	IIIS–1	21 to 24	VA–2	13 to 20
IS–3	28	IIS–2	25 to 32	IIIP–2	25 to 36	VA–3	21
IS–4	29	IIS–3	33	IIIS–2	37	VA–4	22
IS–3	30	IIS–4	34	IIIS–3	38		23ᵃ
IS–5	31, 32	IIP–2	35 to 42	IIIS–4	39	VA–6	24
IS–4	33, 34	IIP–3	43 to 58	IIIP–3	40 to 43	VA–7.1	25
IS–3	35 to 40	IIP–4	59 to 66	IIIP–4	44 to 51	VA–8	26
IS–6	41	IIP–5	67 to 74	IIIP–5	52 to 67	VA–9	27
IS–7	42	IIS–5	75	IIIP–6	68 to 71	VA–10	28
	43ᵃ	IIP–6	76 to 91	IIIS–5	72	VA–11.2	29
IS–6	44	IIS–6	92	IIIS–6	73	VA–12	30
	45ᵃ	IIS–7	93	IIIS–7	74 to 77	VA–13	31
IS–7	46 to 49	IIS–8	94	IIIS–8	78	VA–14	32
IS–6	50, 51	IIS–9	95	IIIS–9	79	V–1	33 to 36
IP–1	52 to 67	IIP–7	96 to 111	IIIS–10	80 to 83	VA–2.1	37
IP–2	68 to 83	IIS–10.2	112	IIIS–11	84	V–3.1	38
IS–8	84	IIP–8	113–136	IIIS–13	85	VA–15	39
IP–3	85 to 100	IIS–11	137	IIIP–7	86 to 101	V–4	40
	101ᵃ	IIP–9	138 to 145	IIIS–14	102	V–5.1	41
IS–9	102	IIP–10	146 to 161	IIIS–15	103 to 106	V–6	42
IS–10	103	IIS–12	162	IIIS–16	107	VA–16.1	43
	104ᵇ	IIP–11	163 to 178	IIIS–17	108 to 111	V–8	44 to 51
IP–4	105 to 112	IIP–12	179 to 194	IIIS–18	112 to 115	V–9.1	52
IS–2	113, 114	IIS–13	195	IIIS–19	116 to 119	VA–17.1	53
IS–9	115, 116	IIS–14	196	IIIS–21	120	V–10	54
IP–5	118 to 133	IIP–13	197 to 212	IIIS–21.5	121	V–11	55 to 62
IS–13	134, 135	IIS–15	213	IIIS–22	122	V–12	63
IS–9	136	IIS–16	214	IIIS–20	123	V–13	64
IS–14	137	IIS–17	215	IIIP–8	124 to 131	VA–18	65
IS–15	138			IIIS–23	132 to 135	V–14	66 to 69
IS–14	139, 140			IIIS–24	136	V–15.1	70
IP–6	141 to 148			IIIP–9	137 to 160	V–16	71 to 78
IS–16	149			IIIS–25	161	VA–19	79
IS–17	150			IIIS–26	162	V–18	80 to 83
IS–16	151			IIIP–10	163 to 170	V–19	84
	152ᵃ			IIIS–27	171	VA–20	85
IS–18	153 to 156			IIIS–28	172	V–21	86 to 89
IP–7	157 to 172			IIIP–11	173 to 180	V–22	90 to 93
IS–19	173			IIIP–12	181 to 212	V–23.1	94 to 97
IS–20	174			IIIS–29	213	V–24	98 to 101
IS–21	175			IIIS–30	214	V–25	102
IP–8.1	176 to 183			IIIS–31	215	VA–21	103
IP–9.2	184 to 215					V–26.1	104 to 107
						V–27	108 to 115
						V–28	116 to 119
						V–29	120 to 123
						VA–22	124
						V–30	125 to 128
						V–31	129 to 132
						V–32	133 to 136
						V–33	137
						V–34	138 to 141
						V–35	142 to 145
						V–36	146 to 149
						V–37	150 to 157
						VA–23	158
						V–38	159 to 162
						VA–24	163
						V–40	164 to 167
						V–41	168
						V–42	169 to 176
						V–43.2	177 to 180
						VA–25	181
						V–45.1	182 to 185
						V–46	186 to 193
						V–48	194 to 201
						V–49	202 to 205
						V–50	206 to 209
						V–51	210 to 217

ᵃ Film-handling considerations required that this frame be advanced through the cameras without being exposed.

ᵇ An exposure taken for diagnostic test purposes. The medium-resolution frame was unexposed; the high-resolution frame was smeared during exposure.

TABLE 8.—*Missions I, II, III, and V Sites for Which Photographs Were Incompletely Read Out or Degraded*[a]

Site	Exposure number	Photo rank[b] Medium-resolution frame	High-resolution frame	Site	Exposure number	Photo rank[b] Medium-resolution frame	High-resolution frame
Mission I				Mission III—Continued			
IS-2	25	A100	C100	IIIP-2	25	A100	NRO
	26, 27	A100	A100		26	NRO	A22
IS-3	28	A100	C100		27	NRO	A1
	30	A100	A100		28	NRO	NRO
	35	A100	C100		29	NRO	NRO
	36	A100	A100		30	NRO	NRO
	37	A100	C100		31	A100	NRO
	38	A100	A100		32	NRO	A19
	39	A100	C100		33	A100	A9
	40	A100	B100		34	NRO	A5
IS-4	29	A100	A100		35	NRO	A22
	33, 34	A100	C100		36	NRO	NRO
IS-5	31	A100	A100	IIIP-3	40	NRO	A65
	32, 34	A100	C100		41	NRO	NRO
IS-6	41	A100	A100		42	NRO	A75
	50, 51	A100	C100		43	NRO	NRO
IS-7	42	A100	A100	IIIP-4	44	NRO	A75
	46 to 49	A100	C100		45	NRO	NRO
IS-9	102, 115	A100	A100		46	NRO	A89
	116	A100	C100		47	NRO	NRO
	117, 136	A100	A100		48	NRO	NRO
Mission II					49	NRO	NRO
					50	A03	NRO
IIP-1	5	A100	NRO		51	NRO	NRO
	6	NRO	A27	IIIP-5	52	NRO	A93
	7	A100	A15		53	A03	NRO
	8	NRO	A36		54	NRO	A96
	9	A100	A23		55	NRO	NRO
	10	NRO	A27		56	NRO	NRO
	11	A100	A27		57	NRO	NRO
	12	A100	NRO		58	A83	NRO
	13 to 20	A100	A100		59	NRO	NRO
IIP-11	163 to 167	A100	A100		60	A100	A71
	168	A50	A100		61	NRO	NRO
	169	A100	A89		62	NRO	A74
	170 to 178	A100	A100		63	A100	NRO
Mission III					64	NRO	A02
					65	NRO	A81
IIIP-1	5	A100	NRO		66	A100	NRO
	6	NRO	A31		67	NRO	A06
	7	NRO	NRO	IIIP-6	68	A100	A84
	8	NRO	NRO		69	NRO	A18
	9	A100	NRO		70	A52	A75
	10	NRO	A74		71	NRO	NRO
	11	A100	NRO	IIIS-1	21	NRO	A11
	12	NRO	A47		22	NRO	NRO
	13	A100	A23		23	NRO	NRO
	14	NRO	A48		24	NRO	NRO
	15	A100	A12	IIIS-2	37	NRO	61[c]
	16	NRO	A51	IIIS-3	38	NRO	NRO
	17	A100	A12	IIIS-4	39	NRO	NRO
	18	NRO	A33	IIIS-5	72	NRO	A77
	19	A100	A13	IIIS-6	73	A86	NRO
	20	NRO	A32	IIIS-7	74	NRO	NRO
					75	NRO	A83
					76	NRO	NRO
					77	NRO	NRO
				Mission V			
				V-51	216	B100[d]	A100
				V-51	217	NRO	A96
				VA-16.1	43	A100	No exp
				VA-18	65	A100	No exp

[a] The photo rank is given for all photographs at each site, but only for those sites where one or more photographs was incompletely read out or degraded. All photographs not listed are ranked A100 except for mission I high-resolution frames which are ranked C100.

[b] Explanation of photo rank. An image quality grade of A, B, or C, based on subjective evaluation, is assigned to each photograph and represents the state of the original film as secured from the spacecraft. This letter is followed by a number expressing the percent of the frame that was read out. Letter grades are: A, a photograph free of image degradation; B, a photograph slightly degraded during exposure in the spacecraft, but which is usable for interpretation; and C, a photograph which was severely degraded during exposure in the spacecraft and which is unusable for interpretation. Consideration is given only to those degradations associated with the operation of the photographic system. Many photographs contain blemishes associated with the spacecraft's development process and others are overexposed to varying degrees. Generally, neither of these seriously affect the usefulness of the photograph for interpretation and are not considered here. NRO indicates the photograph was not read out at all and No exp indicates the spacecraft film was unexposed.

[c] An experimental zero-phase photograph which was appreciably overexposed. It has questionable utility for interpretation.

[d] This photograph was incompletely developed in the spacecraft but is useful for interpretation.

TABLE 9.—*Selenographic Distribution of Mission IV Exposures*

Designation	Latitude at spacecraft nadir	Exposure number for pass—[a]														
		20	21	22	23	24	25	26	27	28	29	30	31	32	33	34
Band G	33° N										165		[b]177		191,192	
Band F	0°			[b]123												
Band N	72° N	[b]110	116	122	128	134	140	[b]146,147	[b]152	158	164	170	[b]176	183	190	
Band D	42° N	109	115	121	127	133	139	145	151	157	163	169	175	182	189	[b]196
Band C	14° N	108	114	120	126	132	138	144	150	156	162	168	174	181	188	[b]195
Band B	14° S	107	113	119	125	131	137	143	149	155	161	167	173	180	187	194
Band A	42° S	106	112	118	124	130	136	142	148	154	160	166	172	179	186	193
Band S	72° S													[b]178	[b]184,185	
Band H	33° S															
Approximate perilune longitude of spacecraft		1° W	7° W	14° W	20° W	27° W	33° W	40° W	46° W	53° W	59° W	66° W	72° W	79° W	85° W	92° W
Approximate apolune longitude of spacecraft				166° E				140° E			121° E		108° E	101° E	95° E	

Designation	Latitude at spacecraft nadir	Exposure number for pass—[a]—Continued													
		6	7	8	9	10	11	12	13	14	15	16	17	18	19
Band G															
Band F		25[b]	[b]30			[b]51				[b]75				[b]99	
Band N		21 to 24[b]	[b]29	[b]36	[b]42	[b]48	[b]56	[b]62	[b]68	[b]74	[b]80	[b]86	92	98	104
Band D		17 to 20[b]	[b]28	[b]35	[b]41	[b]47	[b]55	[b]61	[b]67	[b]73	[b]79	[b]85	91	[b]97	103
Band C		13 to 16[b]	[b]27	[b]34	[b]40	[b]46	[b]54	[b]60	[b]66	[b]72	[b]78	84	[b]90	96	102
Band B		9 to 12[b]	[b]26	[b]33	[b]39	[b]45	[b]53	[b]59	[b]65	[b]71	[b]77	83	[b]89	95	101
Band A		5 to 8[b]		[b]32	[b]38	[b]44	[b]52	[b]58	[b]64	[b]70	[b]76	[b]82	[b]88	[b]94	100
Band S															
Band H															
Approximate perilune longitude of spacecraft		90° E	84° E	77° E	71° E	64° E	58° E	51° E	45° E	38° E	32° E	25° E	19° E	12° E	6° E
Approximate apolune longitude of spacecraft		90° W				116° W				142° W				168° W	

[a] Exposures in latitude bands S, A, B, C, D, and N were taken near perilune with morning illumination. Photographs in bands F, G, and H were taken near apolune with evening illumination.

[b] Exposures for which photographs were incompletely read out or secured in degraded form. See table 11.

TABLE 10.—*Assignment of Mission IV Exposures*

Site[b]	Exposure number(s)	Site[b]	Exposure number(s)	Site[b]	Exposure number(s)	Site[b]	Exposure number(s)	Site[b]	Exposure number(s)	Site[b]	Exposure number(s)
IV-6S	5 to 8	IV-11A	52	IV-16S	82	IV-21A	112	IV-26A	142	IV-31A	172
IV-6A	9 to 12	IV-11B	53	IV-16A	83	IV-21B	113	IV-26B	143	IV-31B	173
IV-6B	13 to 16	IV-11C	54	IV-16B	84	IV-21C	114	IV-26C	144	IV-31C	174
IV-6C	17 to 20	IV-11D	55	IV-16C	85	IV-21D	115	IV-26D	145	IV-31D	175
IV-6D	21 to 24	IV-11N	56	IV-16D	86	IV-21N	116	IV-26F	146	IV-31N	176
IV-6F	25		*57		*87		*117	IV-26F	147	IV-31G	177
IV-7A	26	IV-12S	58	IV-17A	88	IV-22S	118	IV-27A	148	IV-32H	178
IV-7B	27	IV-12A	59	IV-17B	89	IV-22A	119	IV-27B	149	IV-32S	179
IV-7C	28	IV-12B	60	IV-17C	90	IV-22B	120	IV-27C	150	IV-32A	180
IV-7D	29	IV-12C	61	IV-17D	91	IV-22C	121	IV-27D	151	IV-32B	181
IV-7N	30	IV-12D	62	IV-17N	92	IV-22D	122	IV-27N	152	IV-32C	182
	31*		*63		*93	IV-22F	123		*153	IV-32D	183
IV-8S	32	IV-13A	64	IV-18S	94	IV-23A	124	IV-28S	154	IV-33H	184
IV-8A	33	IV-13B	65	IV-18A	95	IV-23B	125	IV-28A	155	IV-33H	185
IV-8B	34	IV-13C	66	IV-18B	96	IV-23C	126	IV-28B	156	IV-33A	186
IV-8C	35	IV-13D	67	IV-18C	97	IV-23D	127	IV-28C	157	IV-33B	187
IV-8D	36	IV-13N	68	IV-18D	98	IV-23N	128	IV-28D	158	IV-33C	188
	37*		*69	IV-18F	99		*129		*159	IV-33D	189
										IV-33N	190
IV-9A	38	IV-14S	70	IV-19A	100					IV-33G	191
IV-9B	39	IV-14A	71	IV-19B	101					IV-33G	192
IV-9C	40	IV-14B	72	IV-19C	102						
IV-9D	41	IV-14C	73	IV-19D	103	IV-24S	130	IV-29A	160	IV-34S	193
IV-9N	42	IV-14D	74	IV-19N	104	IV-24A	131	IV-29B	161	IV-34A	194
	43*	IV-14F	75		*105	IV-24B	132	IV-29C	162	IV-34B	195
						IV-24C	133	IV-29D	163	IV-34C	196
						IV-24D	134	IV-29N	164		
							*135	IV-29G	165		
IV-10S	44	IV-15A	76	IV-20S	106						
IV-10A	45	IV-15B	77	IV-20A	107						
IV-10B	46	IV-15C	78	IV-20B	108						
IV-10C	47	IV-15D	79	IV-20C	109	IV-25A	136	IV-30S	166		
IV-10D	48	IV-15N	80	IV-20D	110	IV-25B	137	IV-30A	167		
	49*		*81		*111	IV-25C	138	IV-30B	168		
	50*					IV-25D	139	IV-30C	169		
IV-10F	51					IV-25N	140	IV-30D	170		
							*141		*171		

* Film-handling considerations required that this frame be advanced through the cameras without being exposed.

[b] Sites designated by: IV denotes mission IV; arabic numeral, pass number; and letters indicate the latitude band of photography. See table 9.

Site	Exposure number	Photo rank [b]		Site	Exposure number	Photo rank [b]	
		Medium-resolution frame	High-resolution frame			Medium-resolution frame	High-resolution frame
IV–6S_____	5	NRO	A100	IV–12S_____	58	B100	B100
	6	NRO	A93	IV–12A_____	59	B100	B100
	7	NRO	NRO	IV–12B_____	60	B100	B100
	8	A100	A59	IV–12C_____	61	C68	B100
IV–6A_____	9	A100	A100	IV–12D_____	62	C100	B100
	10	A100	A93	IV–13A_____	64	C100	B100
	11	A100	A100	IV–13B_____	65	C100	B100
	12	A100	A99	IV–13C_____	66	C100	B100
IV–6B_____	13	No exp	No exp	IV–13D_____	67	C39	B100
	14	No exp	No exp	IV–13N_____	68	C100	B100
	15	No exp	No exp	IV–14S_____	70	C100	A100
	16	No exp	No exp	IV–14A_____	71	B100	B100
IV–6C_____	17	A32	A100	IV–14B_____	72	B100	B100
	18	NRO	A100	IV–14C_____	73	B100	B100
	19	C36	NRO	IV–14D_____	74	B100	A100
	20	NRO	C54	IV–14F_____	75	C100	No exp[c]
IV–6D_____	21	B100	NRO	IV–15A_____	76	B100	A100
	22	NRO	C34	IV–15B_____	77	B100	A100
	23	B100	C01	IV–15C_____	78	B100	A100
	24	NRO	A09	IV–15D_____	79	B100	A100
IV–6F_____	25	C100	No exp[c]	IV–15N_____	80	C100	B100
IV–7A_____	26	No exp	No exp	IV–16S_____	82	B100	A100
IV–7B_____	27	C100	B100	IV–16C_____	85	B50	B100
IV–7C_____	28	C89	C100	IV–16D_____	86	A100	B100
IV–7D_____	29	C100	C100	IV–17A_____	88	B100	A100
IV–7N_____	30	C100	C04	IV–17B_____	89	B100	A100
IV–8S_____	32	C100	C100	IV–17C_____	90	B100	A100
IV–8A_____	33	C100	C100	IV–18S_____	94	B100	A100
IV–8B_____	34	C100	C100	IV–18C_____	97	A89	A100
IV–8C_____	35	C71	C100	IV–18F_____	99	B100	No exp[c]
IV–8D_____	36	C21	C100	IV–20D_____	110	B100	A100
IV–9A_____	38	B100	B100	IV–22F_____	123	B100	No exp[c]
IV–9B_____	39	C100	B100	IV–26F_____	146	A100	No exp[c]
IV–9C_____	40	C100	C100	IV–26F_____	147	A100	No exp[c]
IV–9D_____	41	C64	C100	IV–27N_____	152	A100	A98
IV–9N_____	42	C100	C100	IV–31N_____	176	No exp	A100
IV–10S_____	44	B100	B100	IV–31G_____	177	B100	B100
IV–10A_____	45	C39	B05	IV–32H_____	178	B100	A100
IV–10B_____	46	C100	B100	IV–33H_____	184	B100	A100
IV–10C_____	47	C11	C100	IV–33H_____	185	B100	A100
IV–10D_____	48	C100	C100	IV–34B_____	195	NRO	A100
IV–10F_____	51	B100	NRO	IV–34C_____	196	NRO	A38
IV–11A_____	52	A100	B74				
IV–11B_____	53	B100	B100				
IV–11C_____	54	C68	B100				
IV–11D_____	55	C100	B100				
IV–11N_____	56	C100	B100				

[a] The photo rank is given for all photographs at each site, but only for those sites where one or more photographs were incompletely read out and/or secured in degraded form. All photographs not listed are ranked A100.

[b] Explanation of photo rank. An image quality grade of A, B, or C, based on subjective evaluation, is assigned to each photograph and represents the state of the original film as secured from the spacecraft. This letter is followed by a number expressing the percent of the frame that was read out. Letter grades are: A, a photograph free of image degradation; B, a photograph slightly degraded during exposure in the spacecraft, but which is usable for interpretation; and C, a photograph which was severely degraded during exposure in the spacecraft and which is unusable for interpretation. Consideration is given only to those degradations associated with the operation of the photographic system. Many photographs contain blemishes associated with the spacecraft's development process and others are overexposed to varying degrees. Generally, neither of these blemishes seriously affect the usefulness of the photograph for interpretation and are not considered here. NRO indicates the photograph was not read out at all and No exp indicates the spacecraft film was unexposed.

[c] Mission IV exposures taken at apolune for coverage of the far side are 25, 51, 75, 99, 123, 146, and 147. For each exposure, the high-resolution coverage is situated on the unilluminated side of the evening terminator except for small portions of photographs IV–99H, IV–123H, IV–146H, and IV–147H.

TABLE 12.—*Photographs Processed for Emphasis of Detail in Highlights and Shadows*

Area of interest	Photographs for which SP highlights are available	Photographs for which SP-1 lowlights are available
Structures on bright peak and wall of Petavius	V–33M, V–34M, V–36M	
Details of areas in and around Censorinus	V–63M, V–63H	V–63M, V–63H
Area of Littrow Rilles	V–66M, V–66H	
Bright crater in Sulpicius Gallus Region	V–90H	
Bright walls of Hyginus Rille craters	V–96H, V–97H	
Slopes in vicinity of Alpine Valley	V–102H	V–102H
Slopes in vicinity of Hadley Rille	V–104M, V–105M, V–106M, V–107M	V–104M, V–105M, V–106M, V–107M
Bright walls of crater Alphonsus	V–116H, V–117H, V–118H, V–119H	
Walls of crater near Rima Bode II	V–122H	
Bright walls of crater Tycho	V–125H	
Steep slopes in Rima Plato II Region	V–130M, V–131M, V–132M, V–129H, V–130H, V–131H, V–132H	V–130M, V–131M, V–132M, V–129H, V–130H, V–131H, V–132H
Walls of crater Fra Mauro and other slopes	V–138H	V–138H
Bright walls of crater Copernicus	V–152M	
Scarps of the Imbrian flows	V–160H, V–161H	V–160H, V–161H
Steep slopes in vicinity of Tobias Mayer	V–164M, V–165M, V–166M, V–167M	V–164M, V–165M, V–166M, V–167M
Slopes of Gassendi and adjacent territory	V–177M, V–178M, V–179M, V–180M, V–178H, V–179H	V–177M, V–178M, V–179M, V–180M, V–178H, V–179H
Bright slopes of Jura domes and the terra ridges	V–182M, V–184H, V–185H	
Slopes of rilles and craters in the Harbinger Mountains	V–187H, V–188H, V–189H, V–190H, V–191H, V–192H	V–187H, V–188H, V–189H, V–190H, V–191H, V–192H
Walls of crater Aristarchus	V–198H, V199H, V–200H	V–198H, V–199H, V–200H
Sinuous rille and bright walls of Schröter's Valley	V–204H	V–204H

TABLE 13.—*Some Characteristics of Lunar Orbiter Photographs*

Mission	Typical spacecraft altitude, km	Photo characteristics (average values)						
		Medium-resolution frames			High-resolution frames			
		Photo scale (GRE scale)	Ground resolution, m	Framelet width,[a] km	Photo scale (GRE scale)	Ground resolution, m	Framelet width,[a] km	Reassembly code[c]
Photographs of near side								
Mission I:								
Exposures 5 to 27, 29, 31 to 34, 41, and 42	240	1:420 000	40	7.60	1:55 000	5 to 10	1.0	$N=3$
Other exposures	55	1:96 000	10	1.75	1:12 500	≈40	0.23	$N=3$
Mission II	50	1:87 000	10	1.60	1:11 400	1	0.21	$N=3$
Mission III	55	1:96 000	10	1.75	1:12 500	1	0.23	$N=3$
Mission IV:								
Perilune photos:								
Equatorial regions	2710	1:4 700 000	500	86	1:620 000	60	11	$N=3$
Temperate regions	2940	1:5 100 000	500	93	1:670 000	64	12	$N=3$
Polar regions	3520	1:6 100 000	600	111	1:800 000	76	15	$N=3$
Apolune photographs[b]	5650	1:9 800 000	1000	180	1:1 300 000	120	24	$N=1$
Mission V:								
Extreme value	97	1:169 000	20	3.1	1:22 000	2	0.4	$E=1$
Extreme value	243	1:423 000	40	7.60	1:55 000	5	1.0	$E=1$
Photographs of far side								
Mission I	1375	1:2 400 000	240	43	1:310 000	30	5.7	$N=1$
Mission II	1500	1:2 600 000	260	48	1:340 000	30	6.2	$N=1$
Mission III[b]	1463	1:2 500 000	260	46	1:330 000	30	6.1	$N=1$
Mission IV: Apolune photographs[b]	6150	1:10 700 000	100	195	Not applicable			
Mission V:								
Exposures 5 to 30:								
Extreme value[b]	2548	1:4 400 000	450	81	1:580 000	55	11	$N=3$
Extreme value[b]	5758	1:10 000 000	1000	183	1:1 300 000	125	24	$N=3$
Other exposures:								
Extreme value[b]	1181	1:2 000 000	200	37	1:270 000	30	5	$N=3$
Extreme value[b]	1396	1:2 400 000	240	44	1:320 000	30	6	$N=3$

[a] All Lunar Orbiter photographs are distinguished by faint parallel lines running width-wise. These lines are spaced at approximate 20-mm (0.75-inch) intervals on the 20- by 24-inch sections and provide a convenient rule for measuring the ground distances given.

[b] All photographs on this row were taken obliquely. The values given for photographic characteristics apply only to the nadir point which in most cases were located in an unilluminated area; however, the values given provide a gross characterization of these photographs and are given for a comparison with the other listings and for completeness. Photographs without superscripts are vertical photographs.

[c] See text, page 6, for explanation of reassembly code.

19

TABLE 14.—*Lunar Orbiter Photographs Available From the National Space Science Data Center*

Photo rank	Mission I [a]	Mission II	Mission III	Mission IV	Mission V	Total
Medium-resolution frames						
A100 or B100_____	206	207	152	122	211	898
A (<100) or B (<100)____	0	1	5	3	0	9
Total frames_____	206	208	157	125	211	907
20-inch by 24-inch sections (subtotal)___	206	208	157	125	211	907
High-resolution frames						
A100 or B100_____	14	202	138	132	209	695
A (<100) or B (<100)____	0	7	35	9	1	52
Total frames_____	14	209	173	141	210	747
20-inch by 24-inch sections (subtotal)___	42	615	482	417	630	2186
Medium-resolution frames						
C100 _____	0	0	0	22	0	22
C (<100)_____	0	0	0	10	0	10
Not exposed_____	0	0	0	6	0	6
Not read out_____	0	3	54	9	1	67
Total frames_____	0	3	54	47	1	105
20-inch by 24-inch sections (subtotal)__	0	0	0	32	0	32
High-resolution frames						
C100 _____	192	0	0	12	0	204
C (<100)_____	0	0	0	4	0	4
Not exposed_____	0	0	0	11	2	13
Not read out_____	0	2	38	4	0	44
Total frames_____	192	2	38	31	2	265
20-inch by 24-inch sections (subtotal)__	_____	0	0	41	0	41
20-inch by 24-inch sections (subtotal)__	248	823	639	615	841	3166

[a] Copies of all photographs are available from the NSSDC as 20- by 24-inch sections with the exception of the smeared high-resolution frames of mission I. Copies of these photographs are, however, available as 9½-inch roll film or paper.

References

1. BOWKER, DAVID E.; AND HUGHES, J. KENRICK: Lunar Orbiter Photographic Atlas of the Moon. NASA SP–206, 1970.

2. KOSOFSKY, LEON J.; AND EL-BAZ, FAROUK: The Moon as Viewed by Lunar Orbiter. NASA SP–200, 1970.

3. ANON.: Atlas and Gazetteer of the Near Side of the Moon. NASA SP–241, 1970.

4. BEELER, M.; AND MICHLOVITZ, K.: Lunar Orbiter Photographic Data. Data Users' Note, NSSDC 69–05, NASA Goddard Space Flight Center, June 1969.

5. THE BOEING COMPANY: Lunar Orbiter V—Photography. NASA CR–1094, 1968.

6. THE BOEING COMPANY: Lunar Orbiter V—Photographic Mission Summary. NASA CR–1095, 1968.

7. JAFFE, LEONARD D.: Recent Observations of the Moon by Spacecraft. Space Sciences Reviews, vol. 9, no. 4, c. 1969, pp. 491–616.

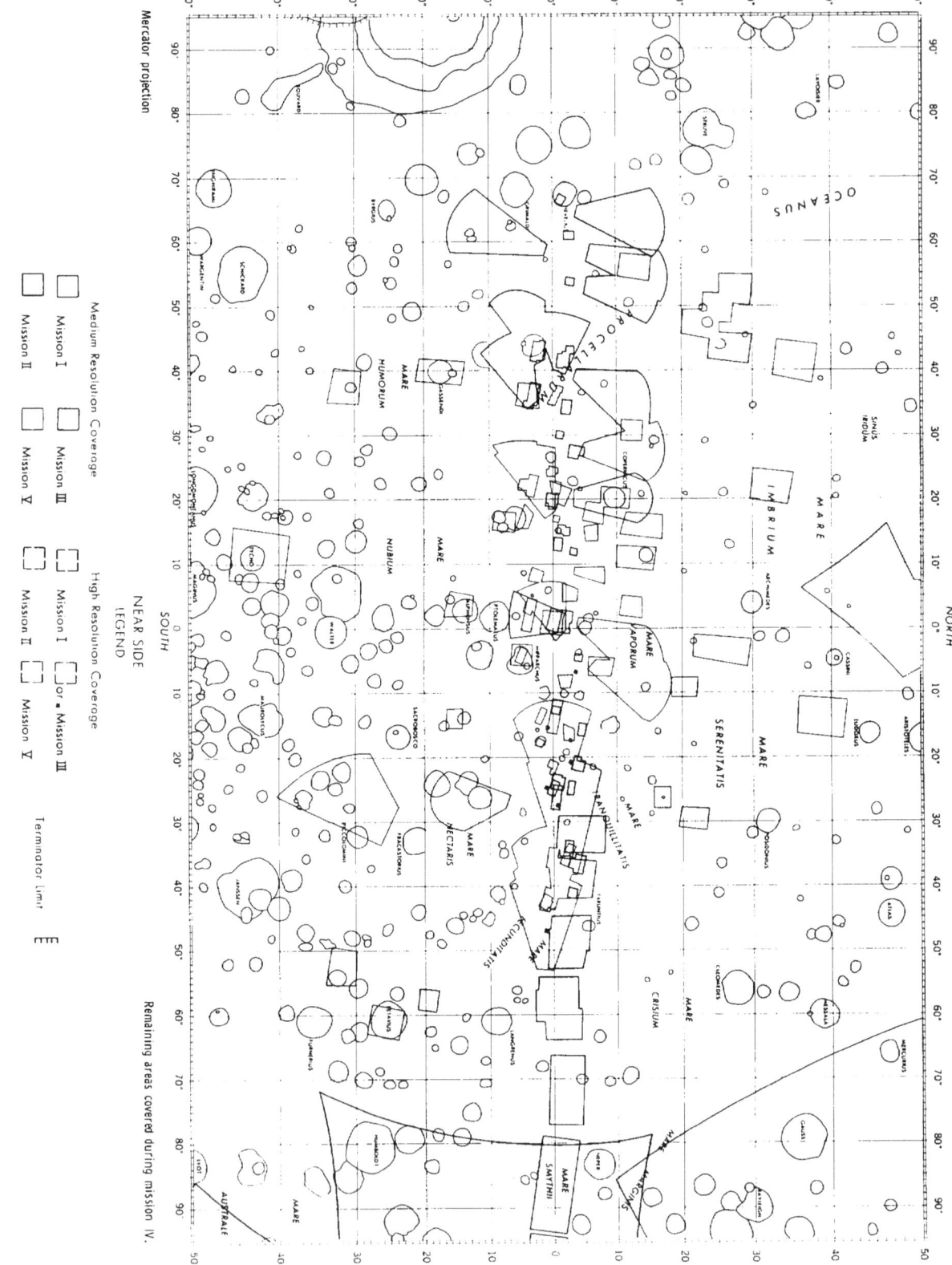

Mercator projection

NEAR SIDE
LEGEND

Medium Resolution Coverage

Mission I   Mission III
Mission II   Mission V

High Resolution Coverage

Mission I   or. Mission III
Mission II   Mission V

Terminator Limit

(a) Equatorial region, near side.

FIGURE 4.—Mission Index for missions I, II, III, and V.

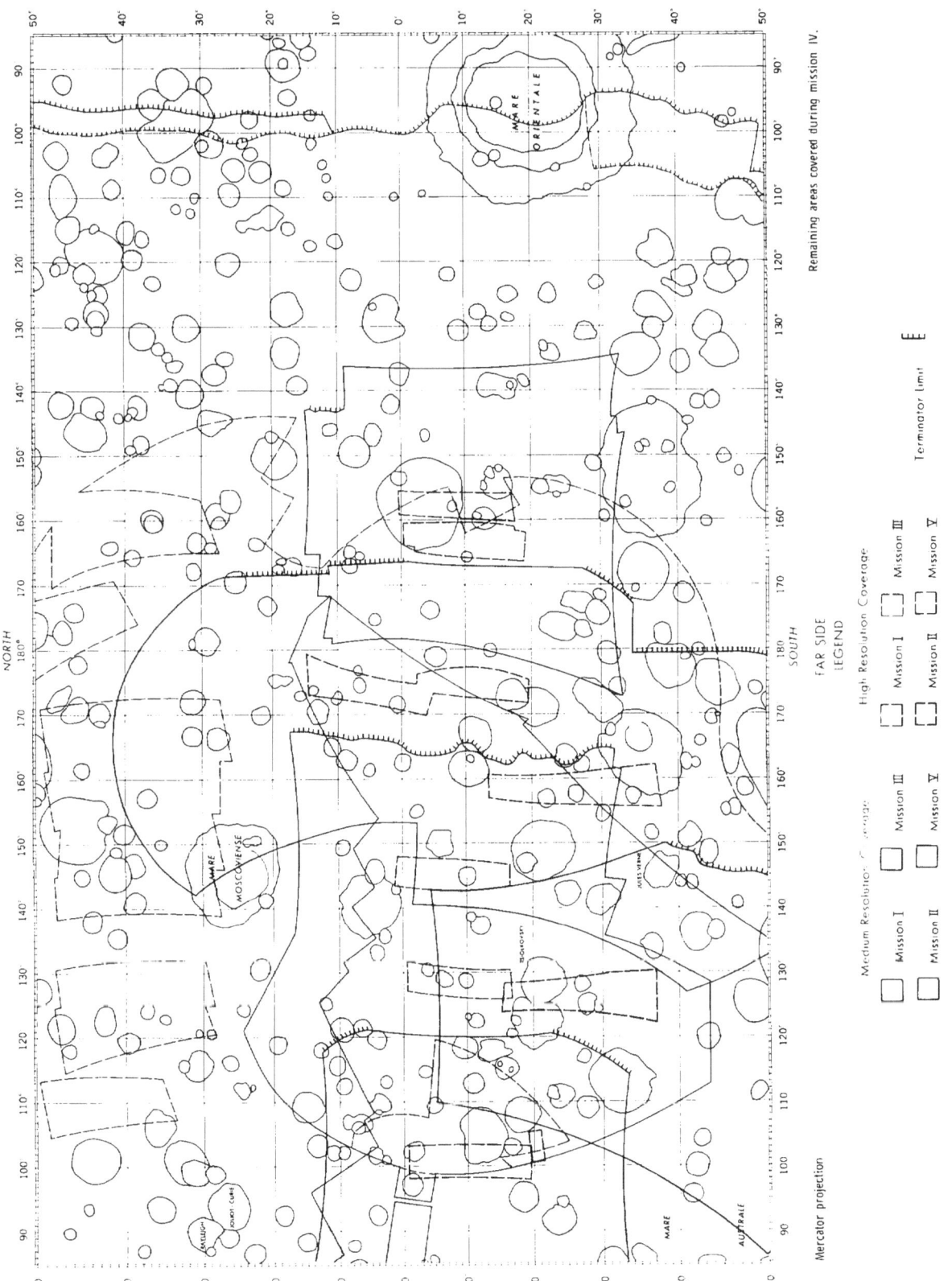

Mercator projection

Remaining areas covered during mission IV.

FAR SIDE

LEGEND

Medium Resolution Coverage

☐ Mission I   ☐ Mission III

☐ Mission II   ☐ Mission V

High Resolution Coverage

☐ Mission I   ☐ Mission III

☐ Mission II   ☐ Mission V

☐ Terminator limit

(b) *Equatorial region, far side.*

FIGURE 4.—*Mission Index for missions 1, 11, 111, and 1*—Continued.

NORTH POLAR REGION

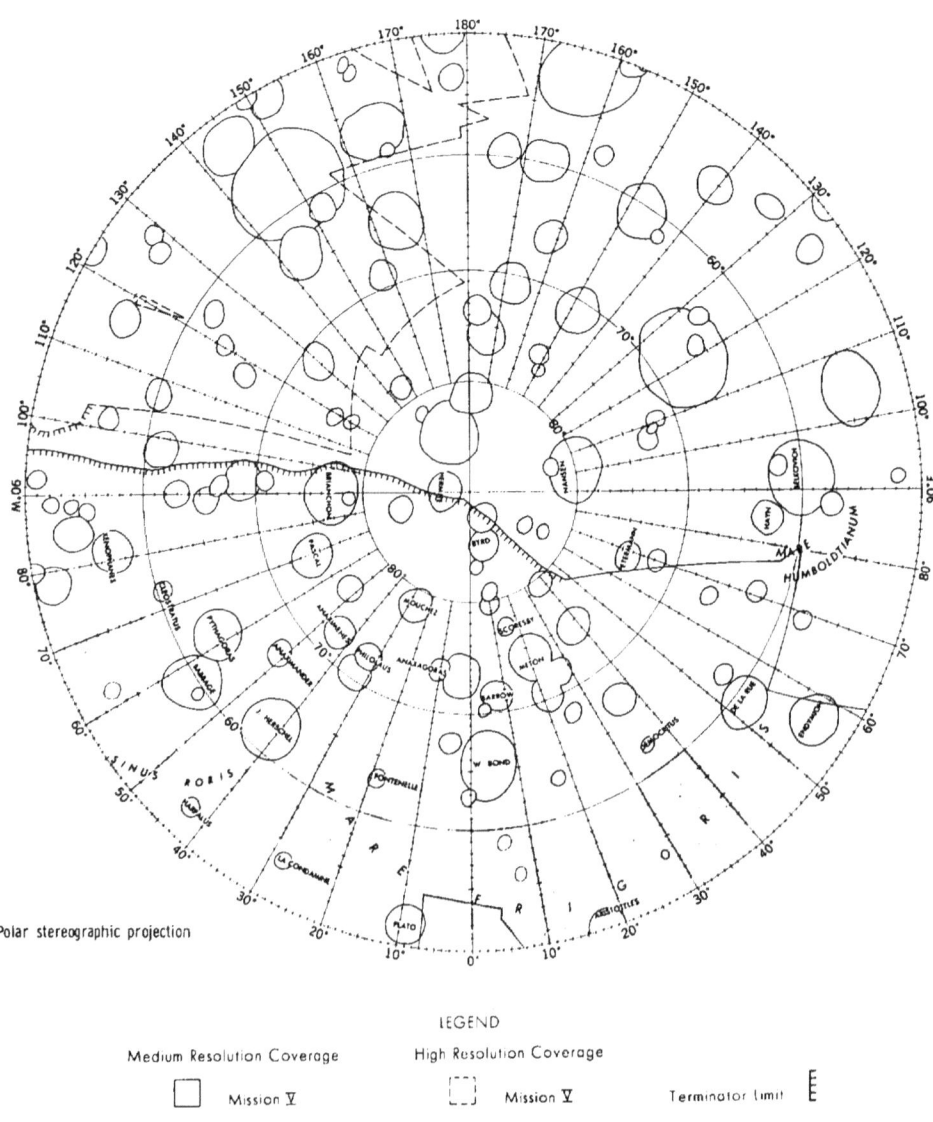

Polar stereographic projection

LEGEND

Medium Resolution Coverage   High Resolution Coverage

Mission V   Mission V   Terminator Limit

Remaining area covered by mission IV.

(c) North polar region.

FIGURE 4.—Mission Index for missions I, II, III, and V.—Continued.

SOUTH POLAR REGION

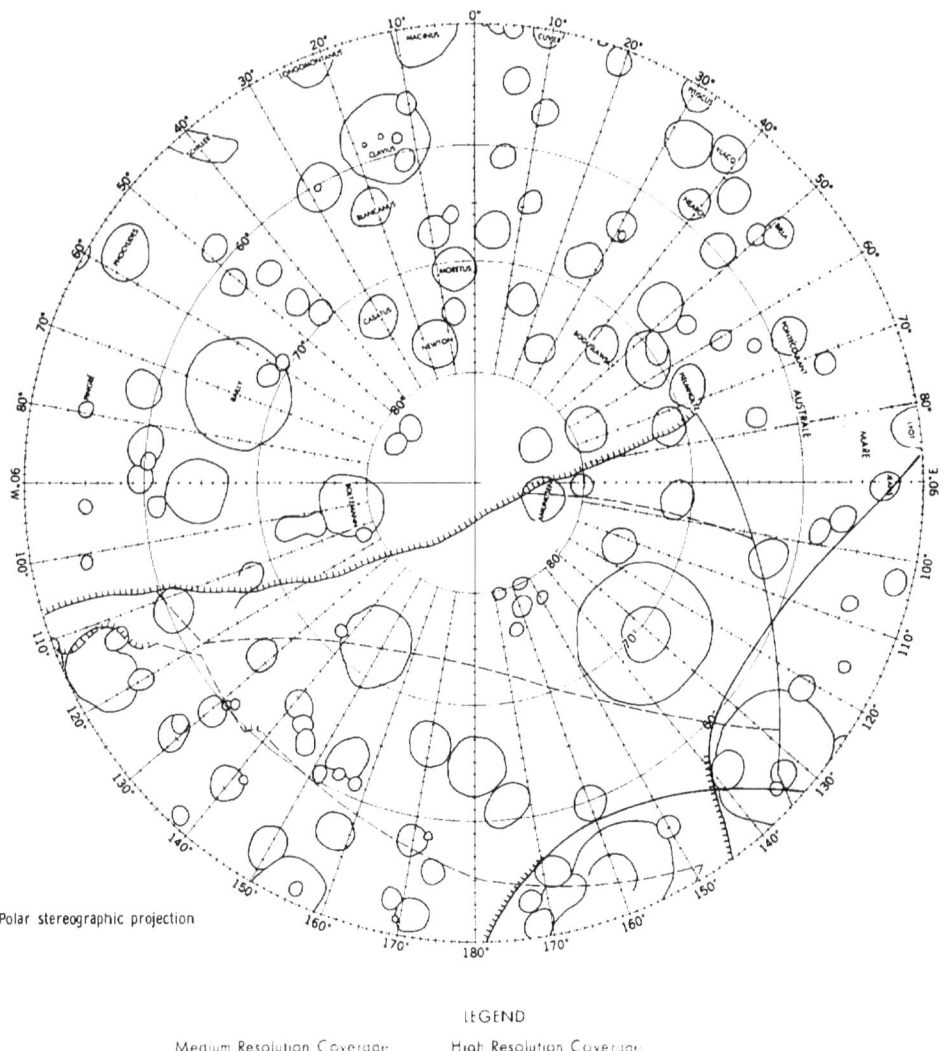

Polar stereographic projection

LEGEND

Medium Resolution Coverage   High Resolution Coverage

Mission II   Mission V   Mission V   Terminator Limit

Mission III

Remaining area covered by mission IV.

(d) South polar region.

FIGURE 4.—Mission Index for missions I, II, III, and V.—Concluded.

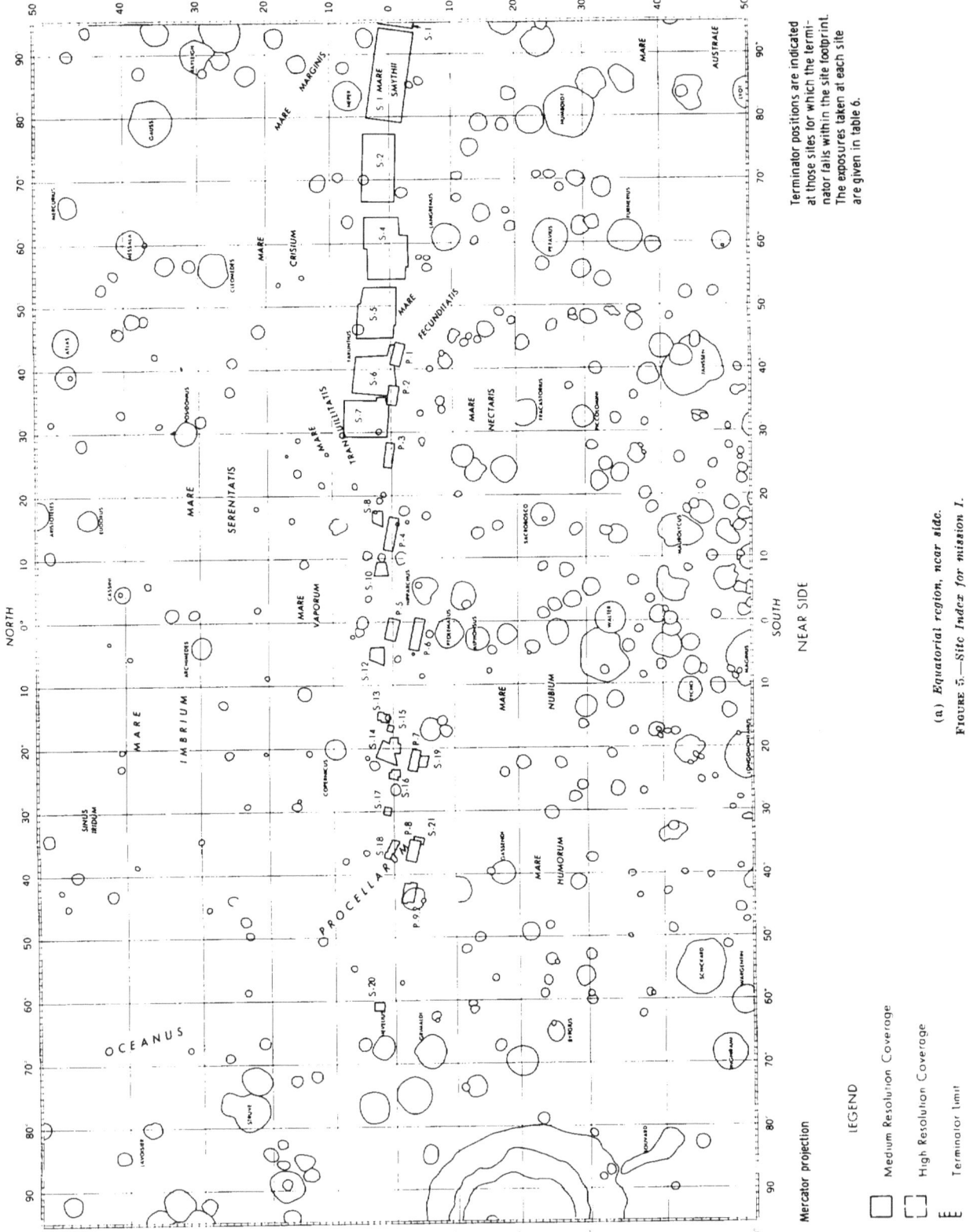

Mercator projection

LEGEND

☐ Medium Resolution Coverage

☐ High Resolution Coverage

Ⲉ Terminator limit

Terminator positions are indicated at those sites for which the terminator falls within the site footprint. The exposures taken at each site are given in table 6.

(a) *Equatorial region, near side.*

FIGURE 5.—*Site Index for mission I.*

26

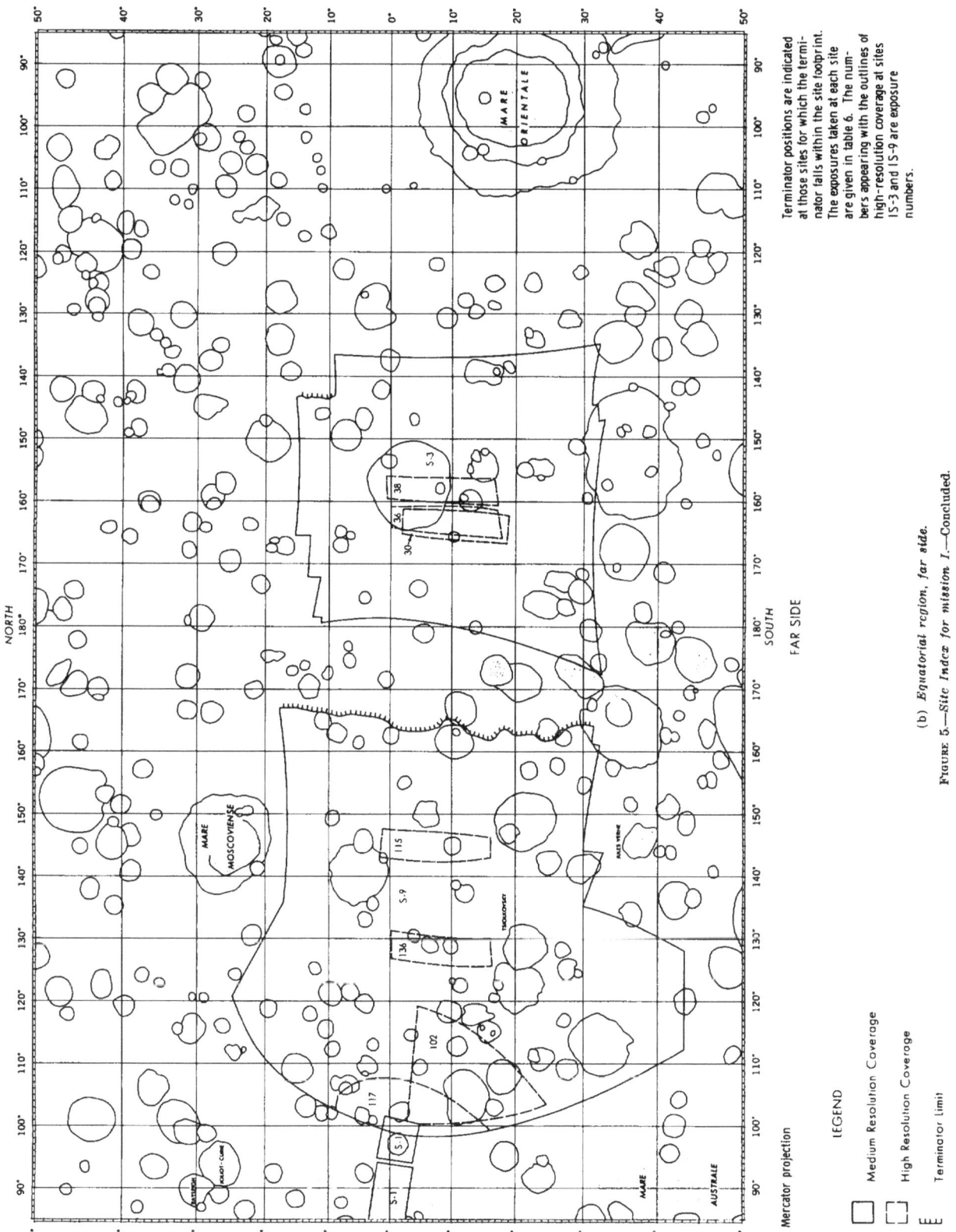

Terminator positions are indicated at those sites for which the terminator falls within the site footprint. The exposures taken at each site are given in table 6. The numbers appearing with the outlines of high-resolution coverage at sites IS-3 and IS-9 are exposure numbers.

(b) *Equatorial region, far side.*

FIGURE 5.—*Site Index for mission I.*—Concluded.

27

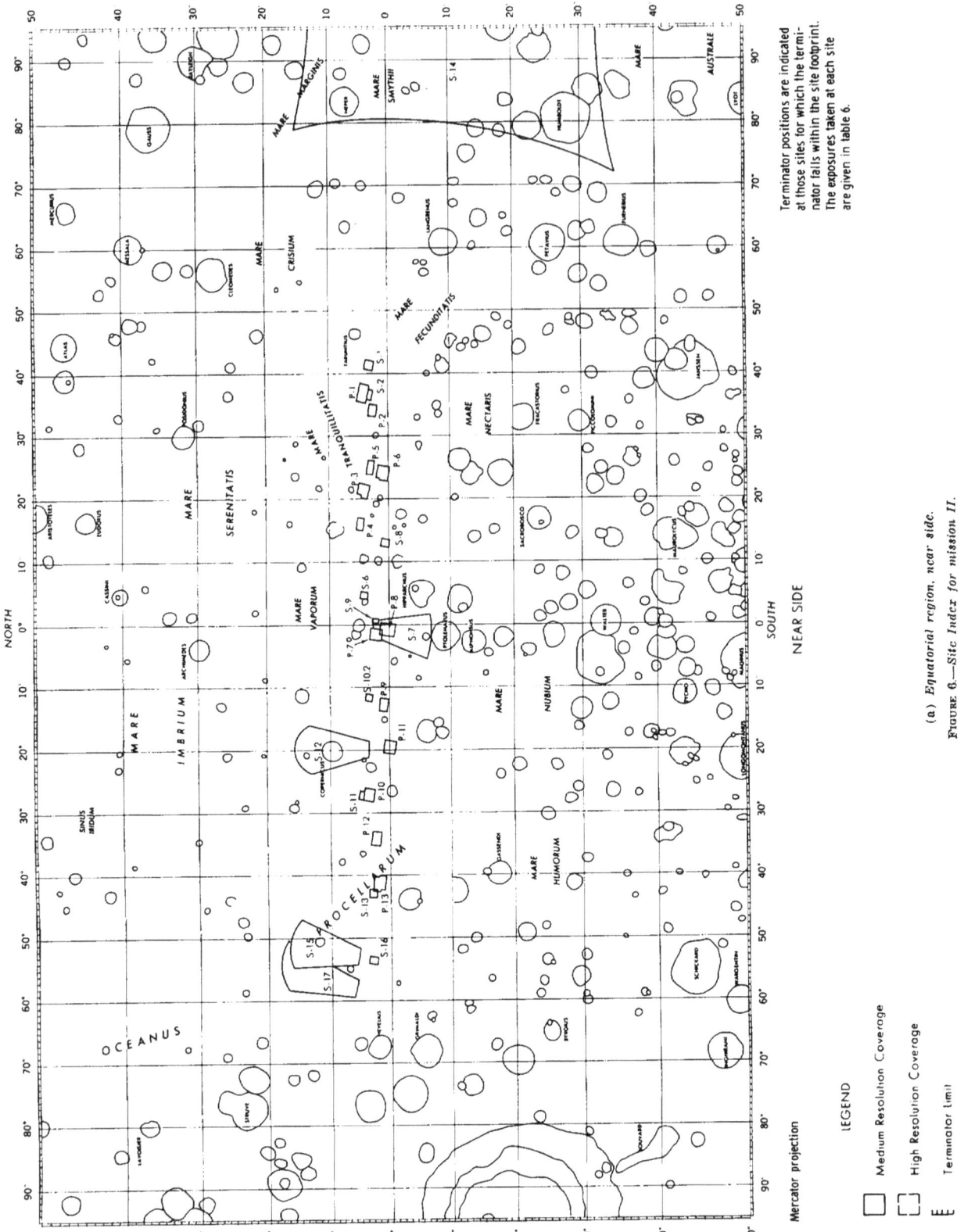

Mercator projection

LEGEND

☐ Medium Resolution Coverage

☐ High Resolution Coverage

E Terminator limit

Terminator positions are indicated at those sites for which the terminator falls within the site footprint. The exposures taken at each site are given in table 6.

(a) *Equatorial region, near side.*

FIGURE 6.—*Site Index for mission II.*

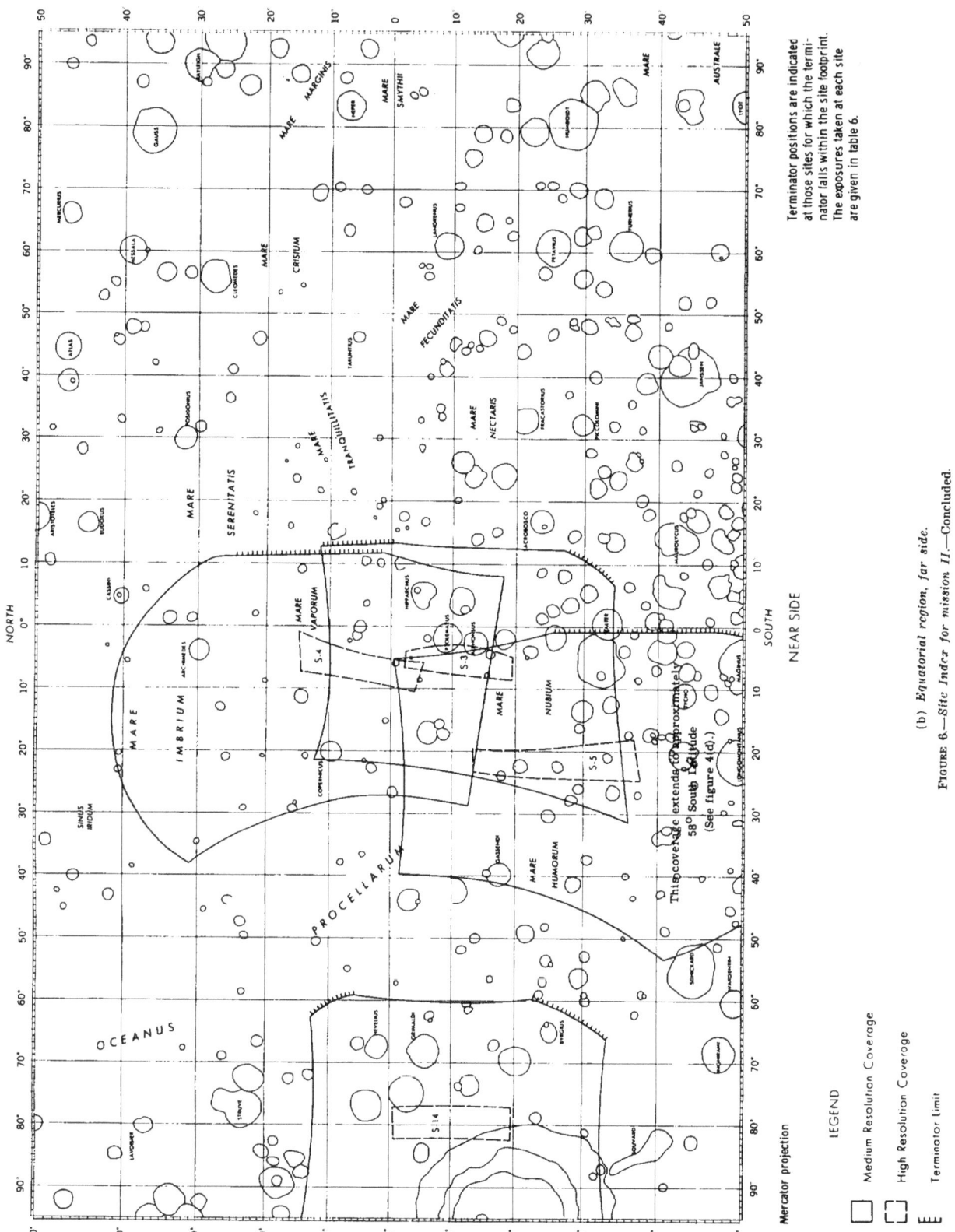

Mercator projection

LEGEND

☐ Medium Resolution Coverage

⬚ High Resolution Coverage

⊞ Terminator Limit

Terminator positions are indicated at those sites for which the terminator falls within the site footprint. The exposures taken at each site are given in table 6.

(b) *Equatorial region, far side.*

FIGURE 6.—*Site Index for mission II.*—Concluded.

29

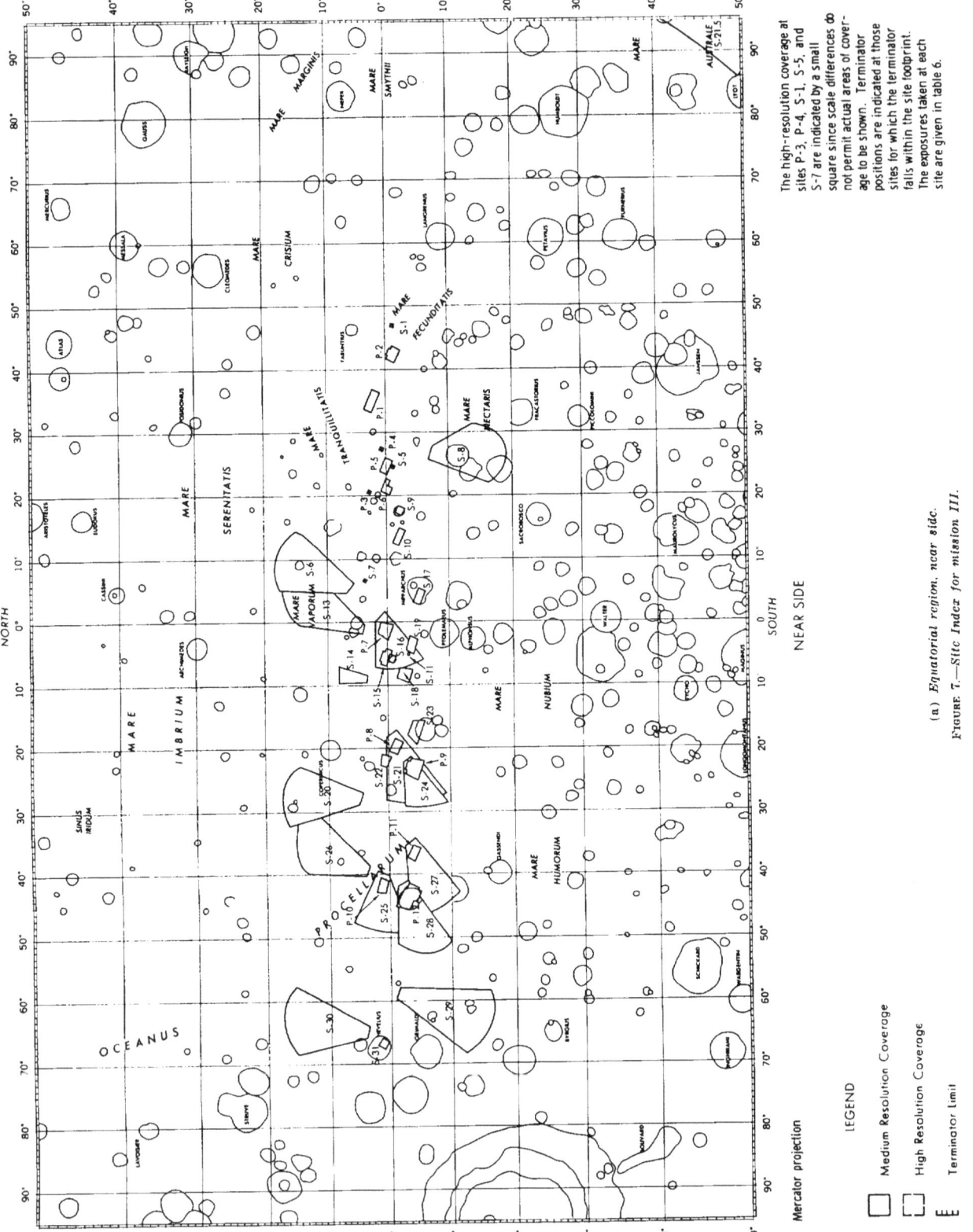

The high-resolution coverage at sites P-3, P-4, S-1, S-5, and S-7 are indicated by a small square since scale differences do not permit actual areas of coverage to be shown. Terminator positions are indicated at those sites for which the terminator falls within the site footprint. The exposures taken at each site are given in table 6.

(a) Equatorial region, near side.

FIGURE 7.—Site Index for mission III.

LEGEND

Mercator projection

☐ Medium Resolution Coverage

⬚ High Resolution Coverage

E Terminator Limit

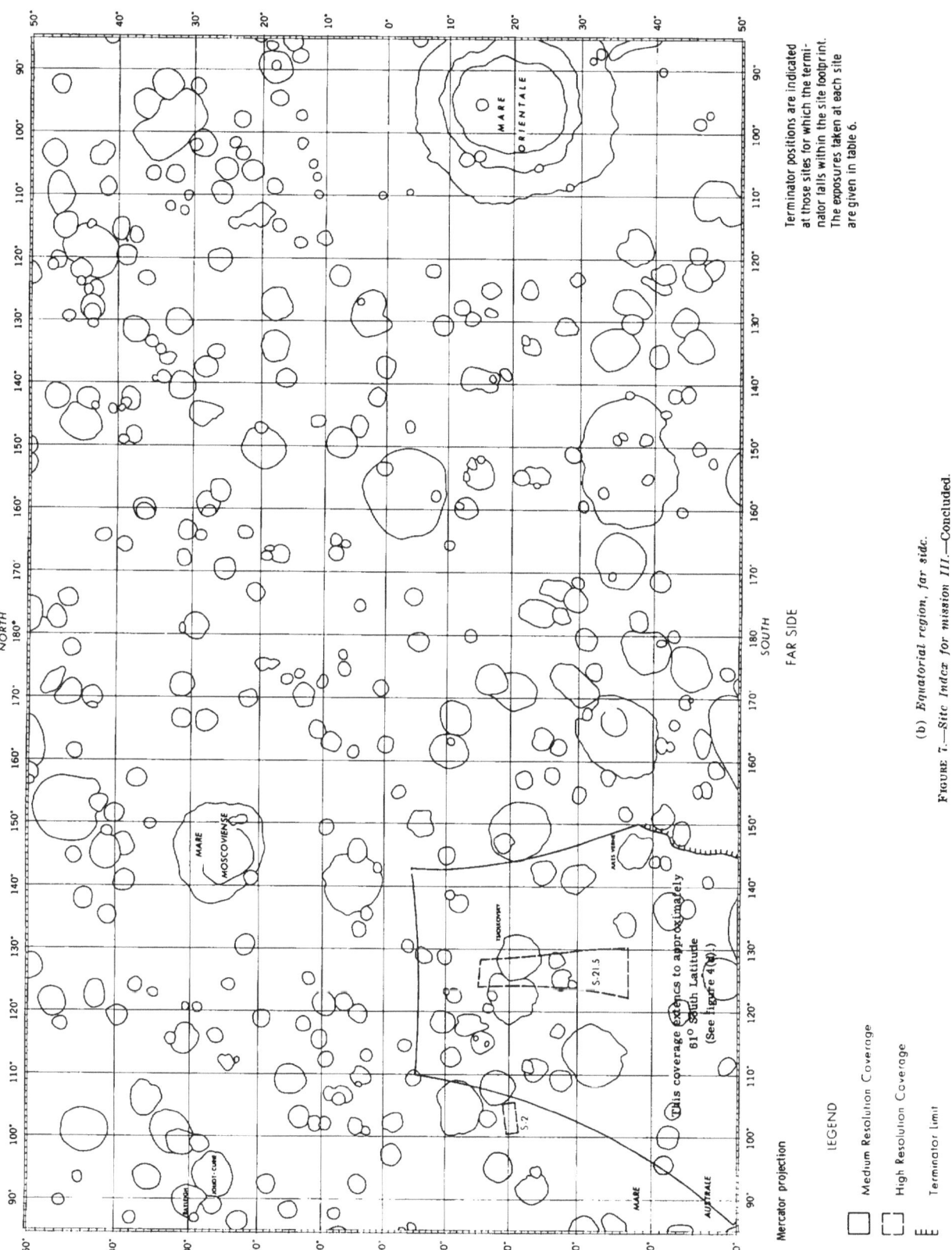

Terminator positions are indicated at those sites for which the terminator falls within the site footprint. The exposures taken at each site are given in table 6.

MARE ORIENTALE

NORTH

FAR SIDE

SOUTH

MARE MOSCOVIENSE

TSIOLKOVSKY

MARE AUSTRALE

This coverage extends to approximately 61° South Latitude (See figure 4(b).)

Mercator projection

LEGEND

☐ Medium Resolution Coverage

☐ High Resolution Coverage

E Terminator limit

(b) *Equatorial region, far side.*

FIGURE 7.—*Site Index for mission III.*—Concluded.

31

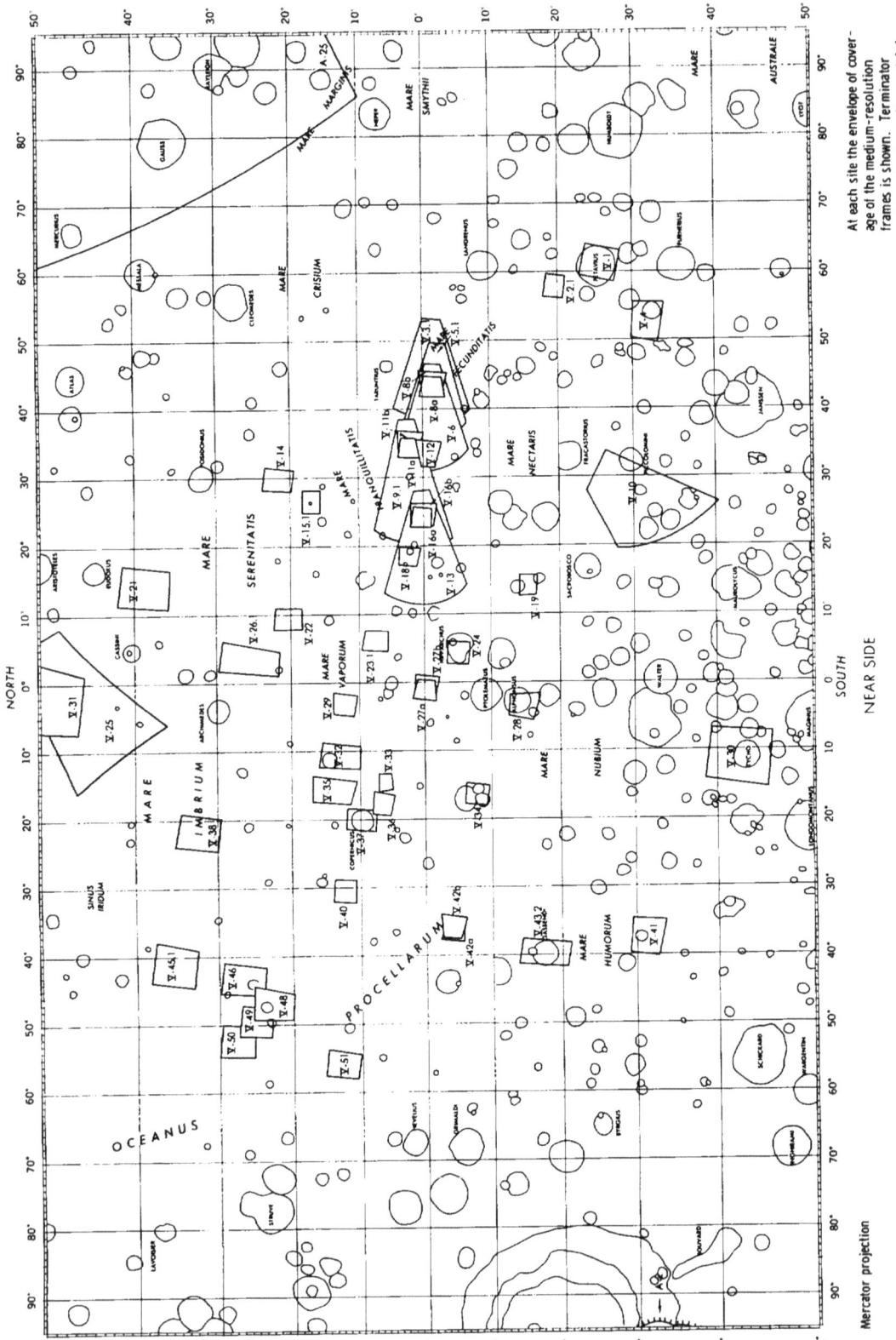

At each site the envelope of coverage of the medium-resolution frames is shown. Terminator positions for those sites containing the terminator are indicated by delineating ticks. The exposures taken at each site are given in table 6.

(a) *Equatorial region, near side.*

FIGURE 8.—*Site Index for mission V.*

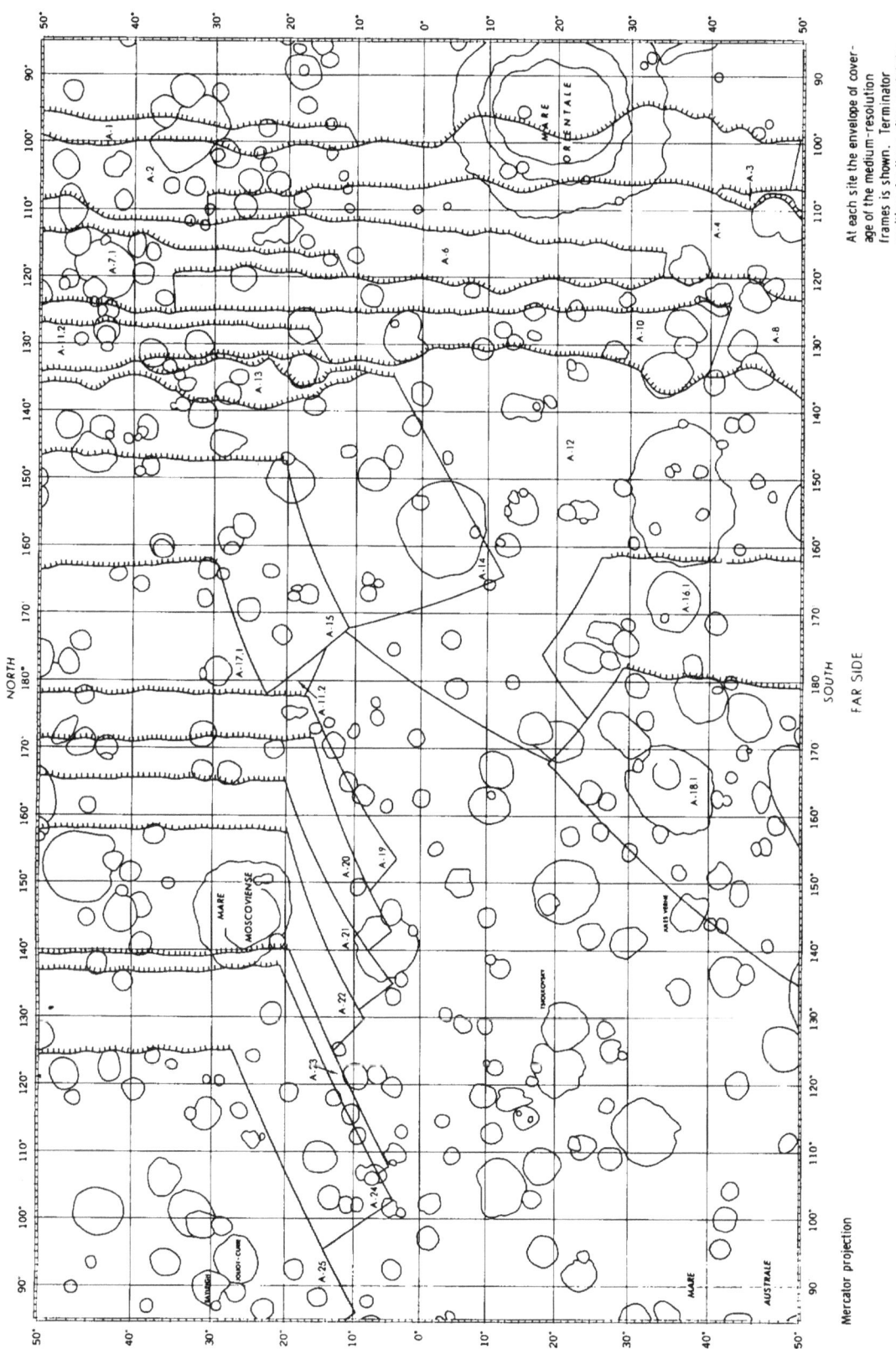

At each site the envelope of coverage of the medium-resolution
frames is shown. Terminator
positions for those sites containing
the terminator are indicated by
delineating ticks. The exposures
taken at each site are given in
table 6.

(b) *Equatorial region, far side.*

FIGURE 8.—*Site Index for mission V.*—Continued.

33

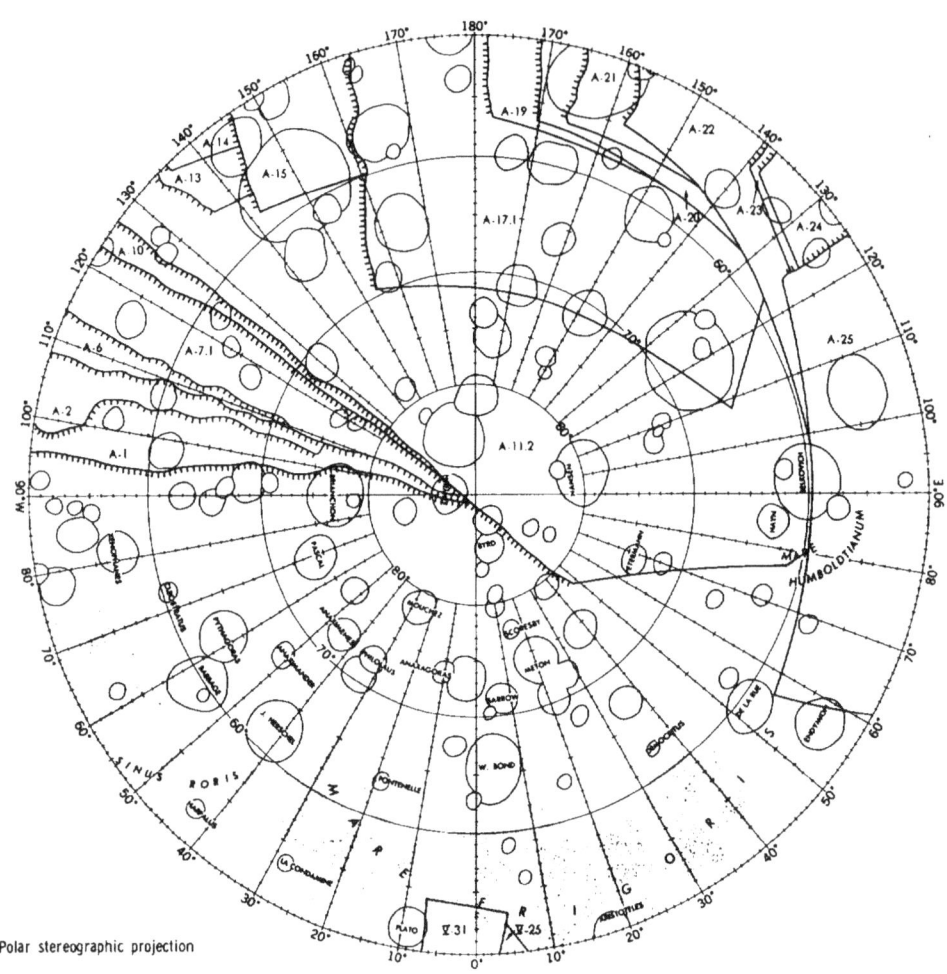

Polar stereographic projection

At each site the envelope of coverage of the medium-resolution
frames is shown. Terminator
positions for those sites containing
the terminator are indicated by
delineating ticks. The exposures
taken at each site are given in
table 6.

(c) *North polar region.*

FIGURE 8.—*Site Index for mission V.*—Continued.

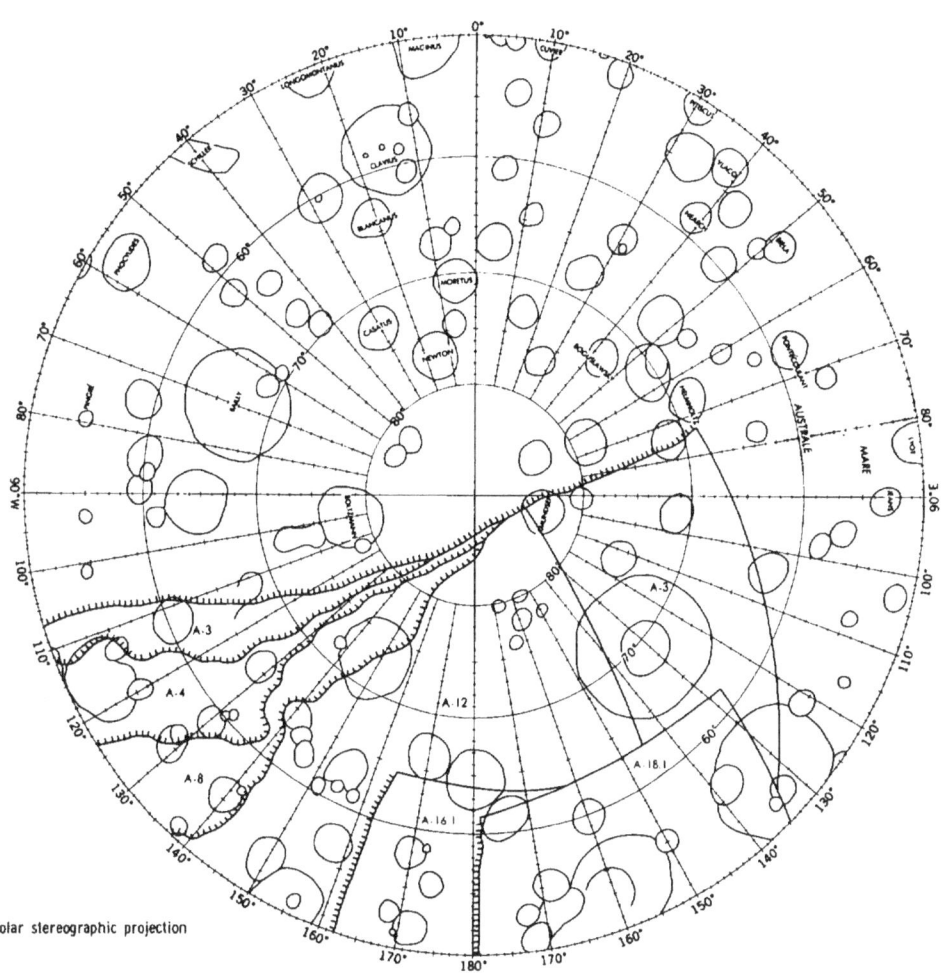

Polar stereographic projection

At each site the envelope of coverage of the medium-resolution frames is shown. Terminator positions for those sites containing the terminator are indicated by delineating ticks. The exposures taken at each site are given in table 6.

(d) *South polar region.*

FIGURE 8.—*Site Index for mission V.*—Concluded.

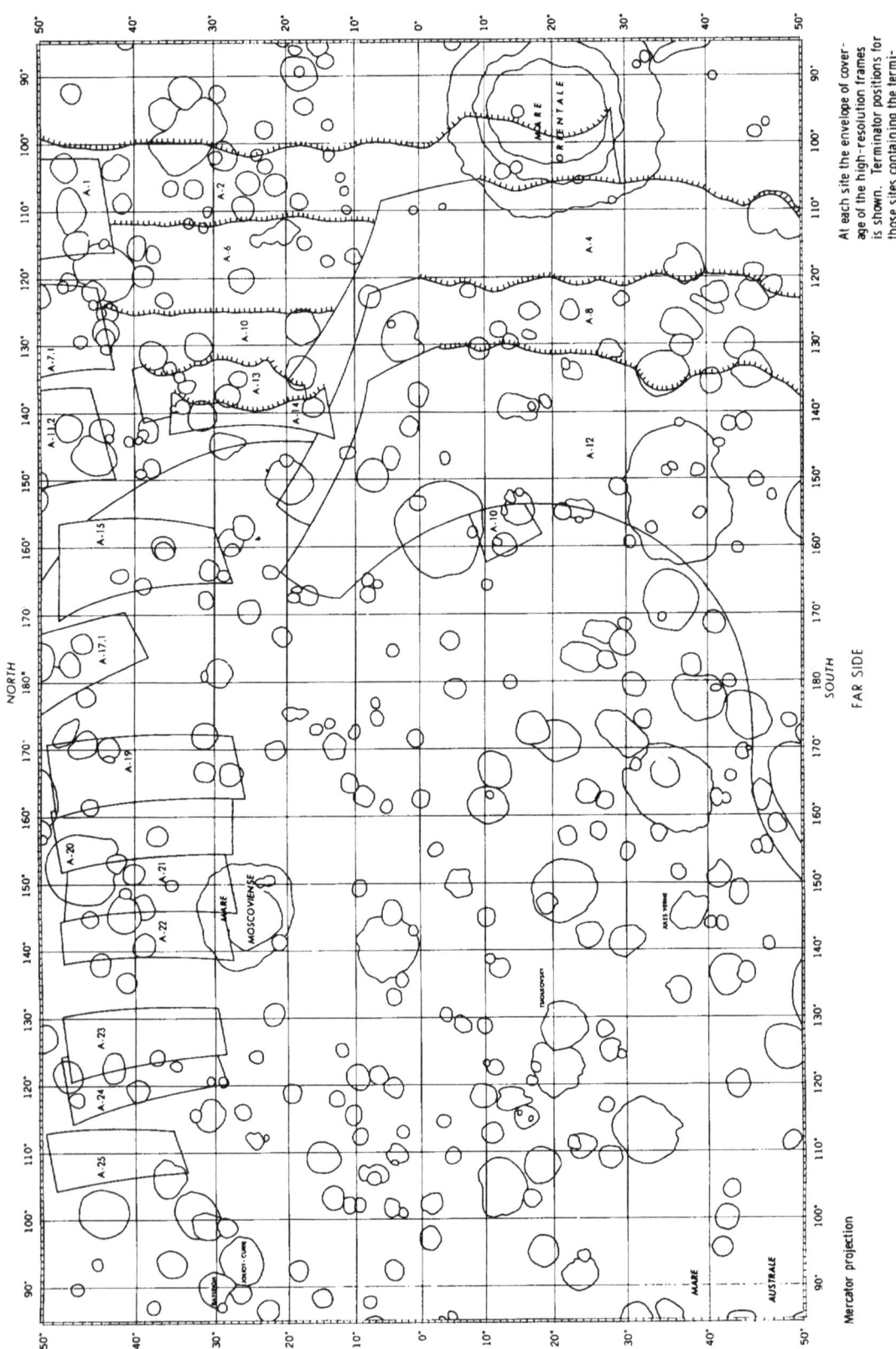

At each site the envelope of coverage of the high-resolution frames is shown. Terminator positions for those sites containing the terminator are indicated by delineating ticks. The exposures taken at each site are given in table 6.

(a) *Equatorial region, far side.*

FIGURE 9.—*Photographic Indexes for mission V high-resolution frames of the far side.*

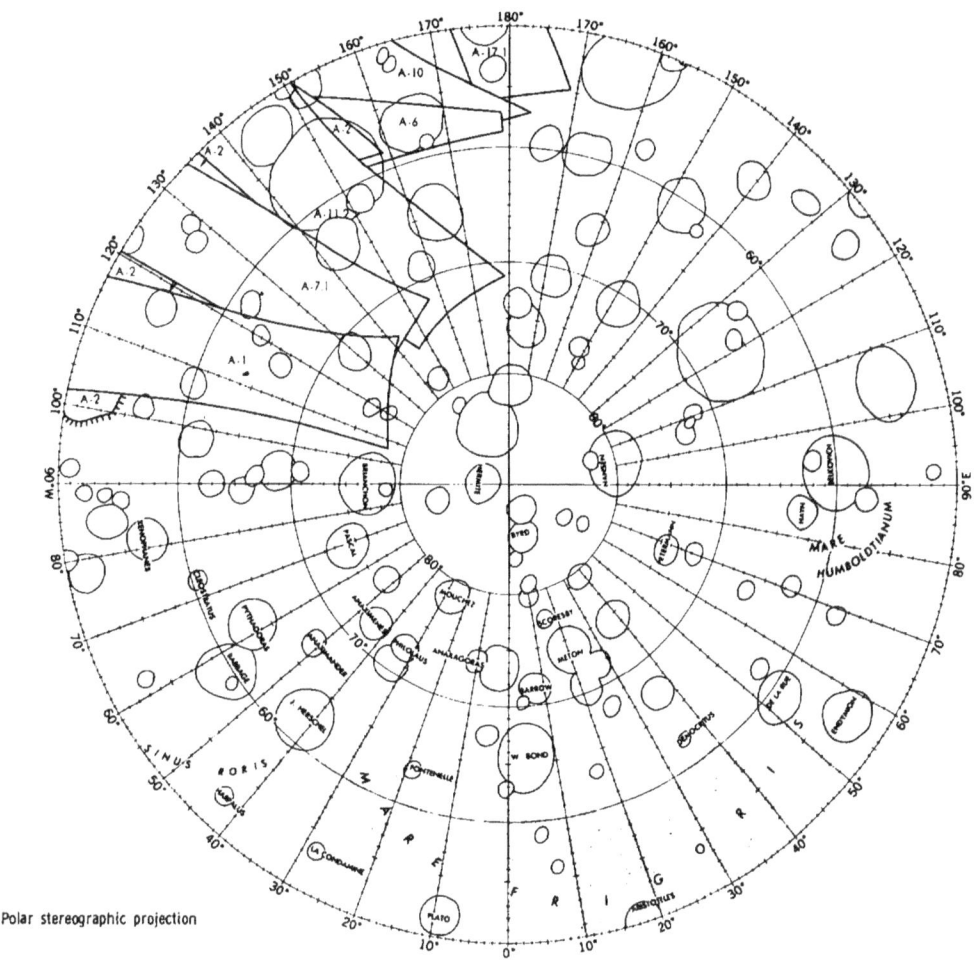

Polar stereographic projection

At each site the envelope of coverage of the high-resolution frames is shown. Terminator positions for those sites containing the terminator are indicated by delineating ticks. The exposures taken at each site are given in table 6.

(b) *North polar region.*

FIGURE 9.—*Photographic Indexes for mission V high-resolution frames of the far side.*—Continued.

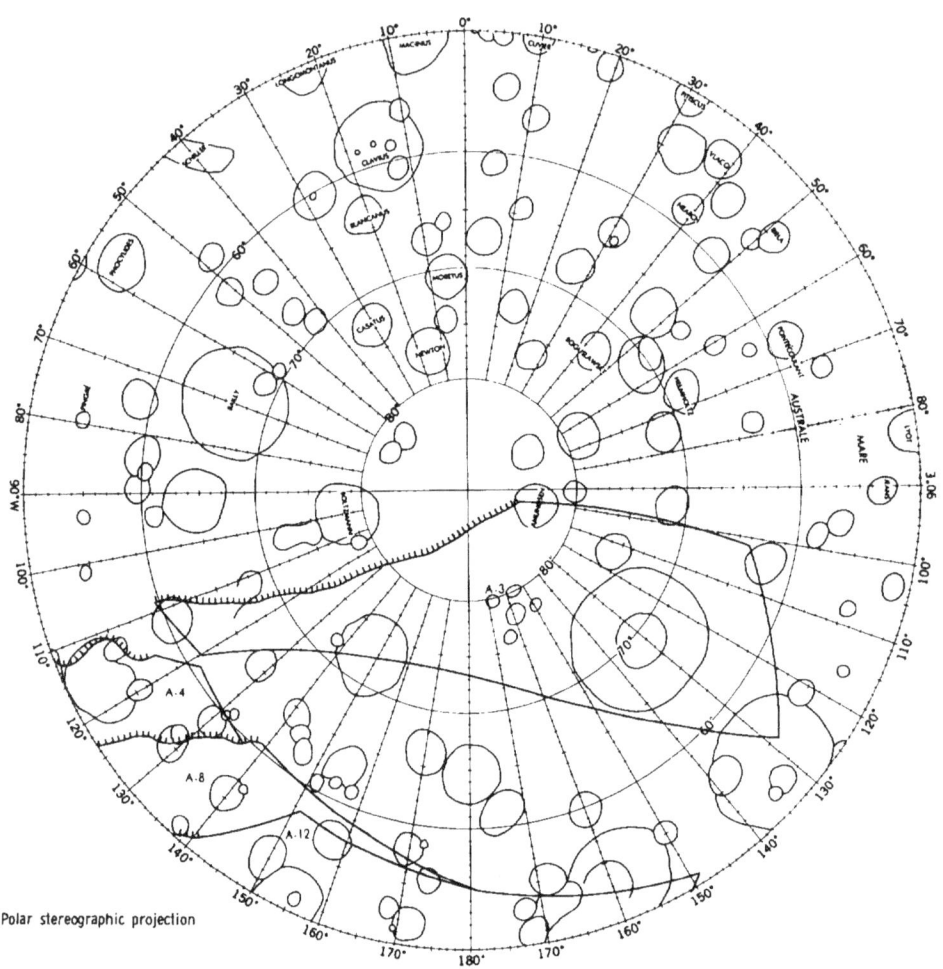

Polar stereographic projection

At each site the envelope of coverage of the high-resolution frames
is shown. Terminator positions for
those sites containing the terminator are indicated by delineating
ticks. The exposures taken at
each site are given in table 6.

(c) *South polar region.*

FIGURE 9.—*Photographic Indexes for mission V high-resolution frames of the far side.*—Concluded.

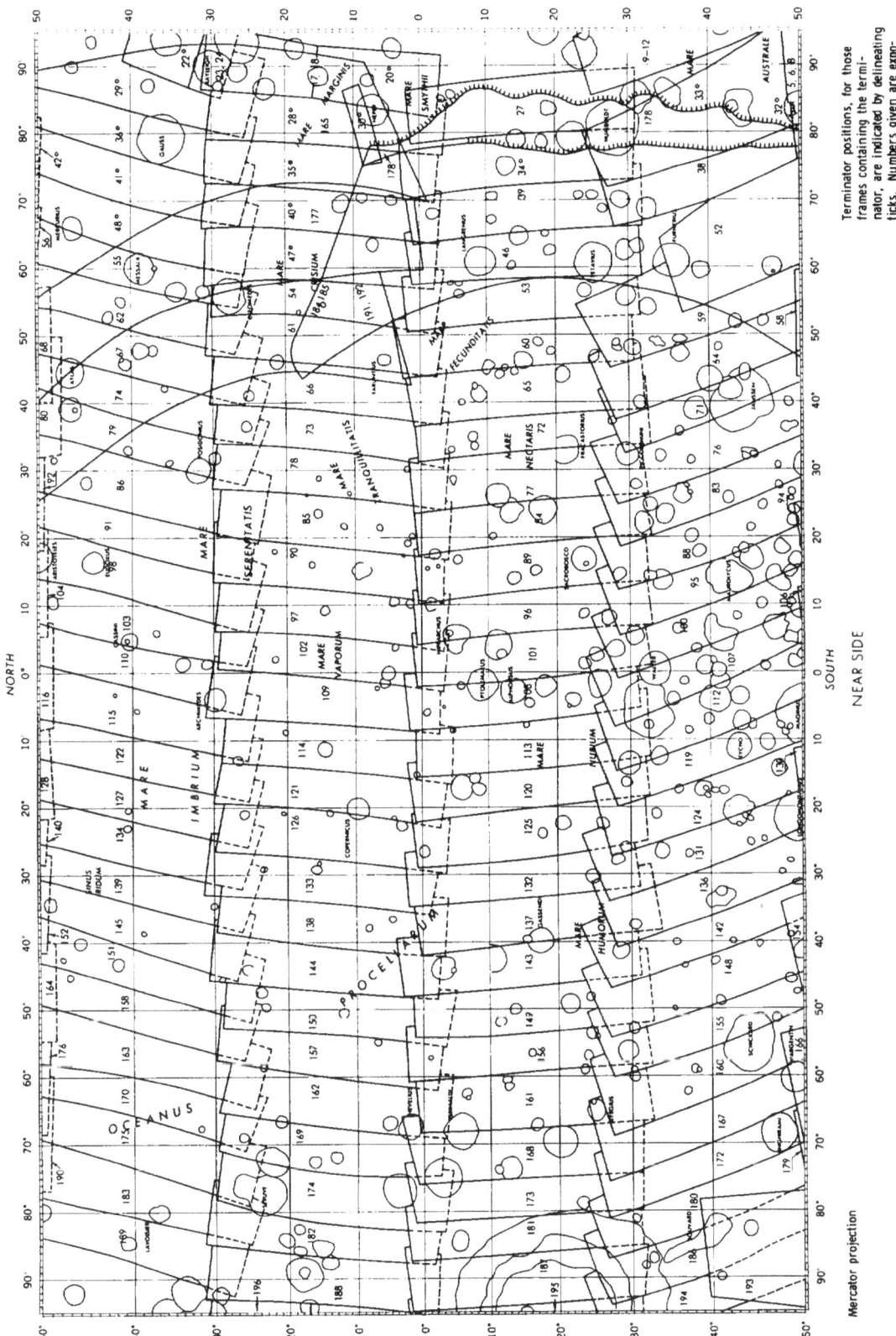

Terminator positions, for those frames containing the terminator, are indicated by delineating ticks. Numbers given are exposure numbers. An asterisk indicates the high-resolution frame is significantly degraded.

(a) Equatorial region, near side.

FIGURE 10.—Photographic Indexes for mission IV high-resolution frames.

39

40

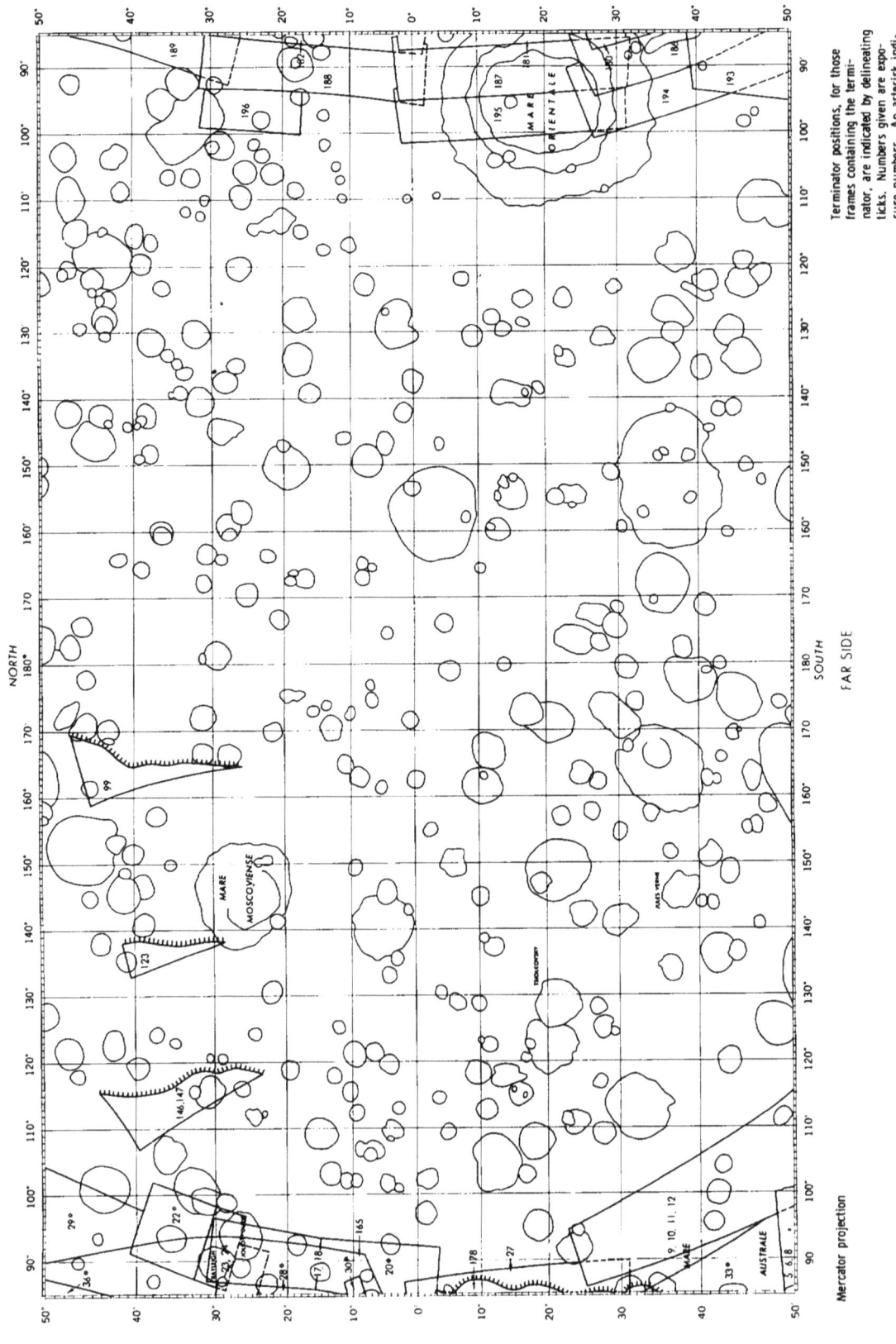

Mercator projection

NORTH

FAR SIDE

SOUTH

Terminator positions, for those frames containing the terminator, are indicated by delineating ticks. Numbers given are exposure numbers. An asterisk indicates the high-resolution frame is significantly degraded.

(b) Equatorial region, far side.

FIGURE 10.—Photographic Indexes for mission II' high-resolution frames.—Continued.

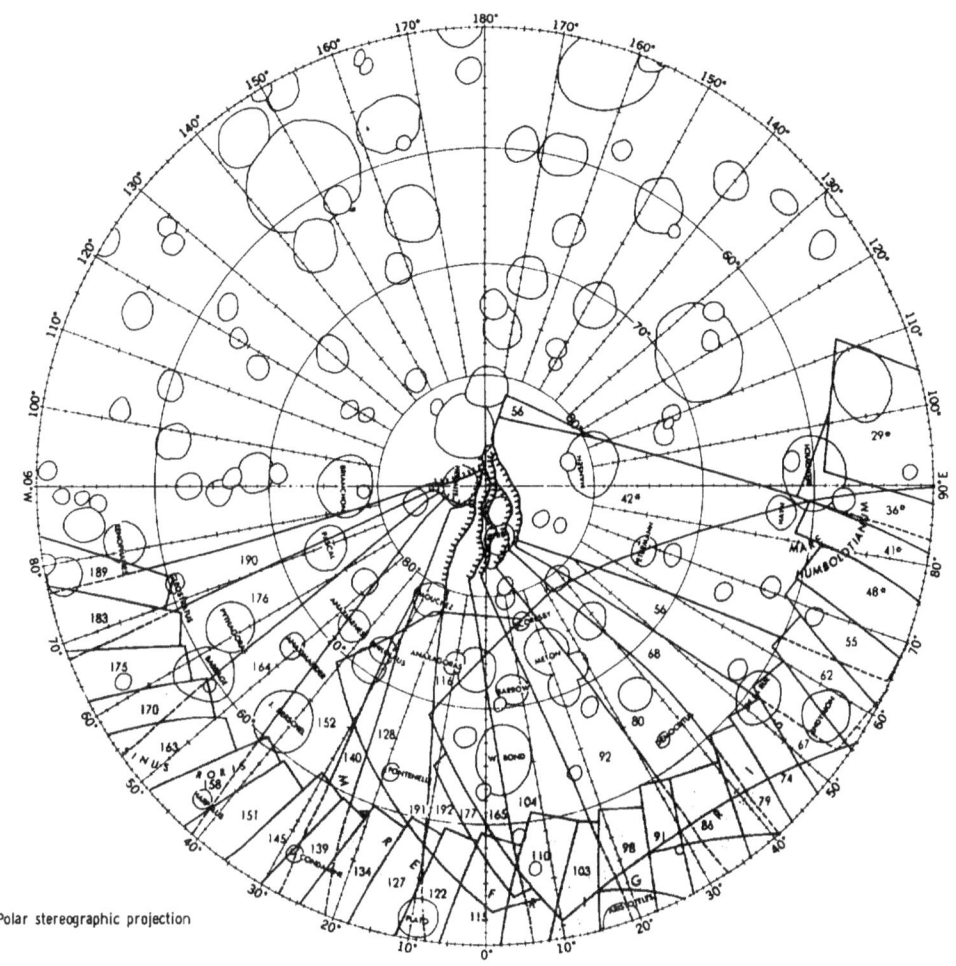

Polar stereographic projection

Terminator positions, for those
frames containing the terminator, are indicated by delineating
ticks. Numbers given are exposure numbers. An asterisk indicates the high-resolution frame
is significantly degraded.

(c) *North polar region.*

FIGURE 10.—*Photographic Indexes for mission IV high-resolution frames.*—Continued.

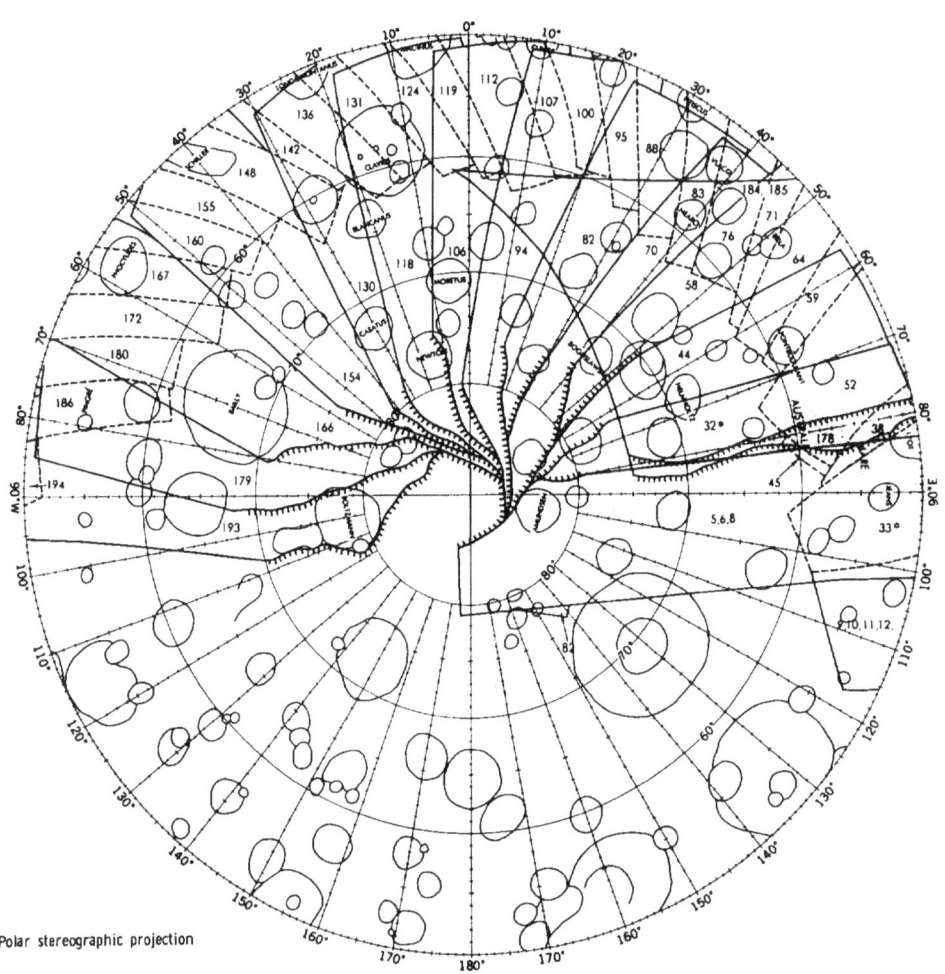

Polar stereographic projection

Terminator positions, for those
frames containing the terminator, are indicated by delineating
ticks. Numbers given are exposure numbers. An asterisk indicates the high-resolution frame
is significantly degraded.

(d) *South polar region.*

FIGURE 10.—*Photographic Indexes for mission IV high-resolution frames.*—Concluded.

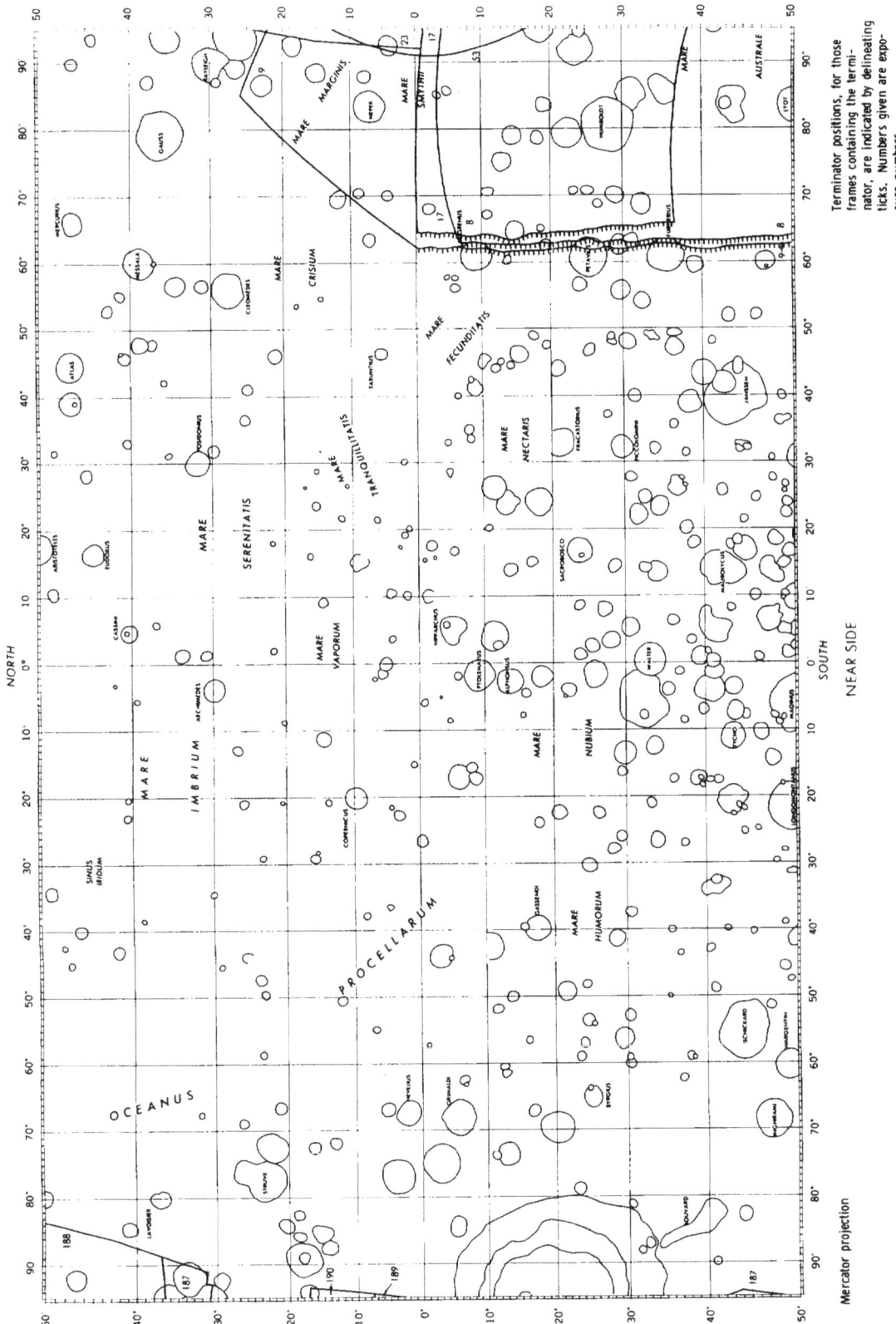

Terminator positions, for those frames containing the terminator, are indicated by delineating ticks. Numbers given are exposure numbers.

(a) *Equatorial region, near side.*

FIGURE 11.—*Photographic Indexes for selected mission IV medium-resolution frames of the far side.*

43

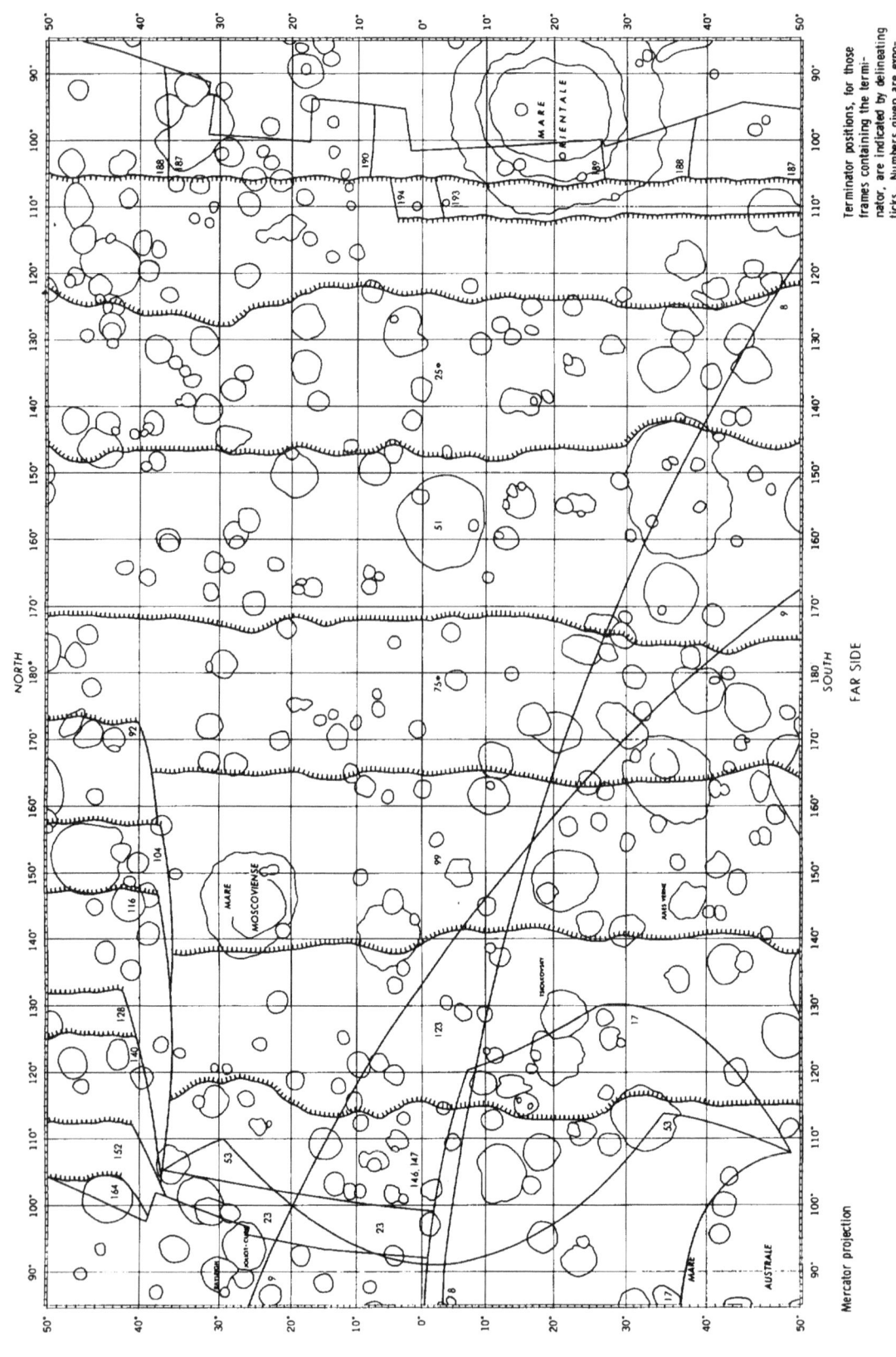

Terminator positions, for those frames containing the terminator, are indicated by delineating ticks. Numbers given are exposure numbers. An asterisk indicates the medium-resolution frame is significantly degraded.

(b) *Equatorial region, far side.*

FIGURE 11.—*Photographic indexes for selected mission IV medium-resolution frames of the far side.*—Continued.

44

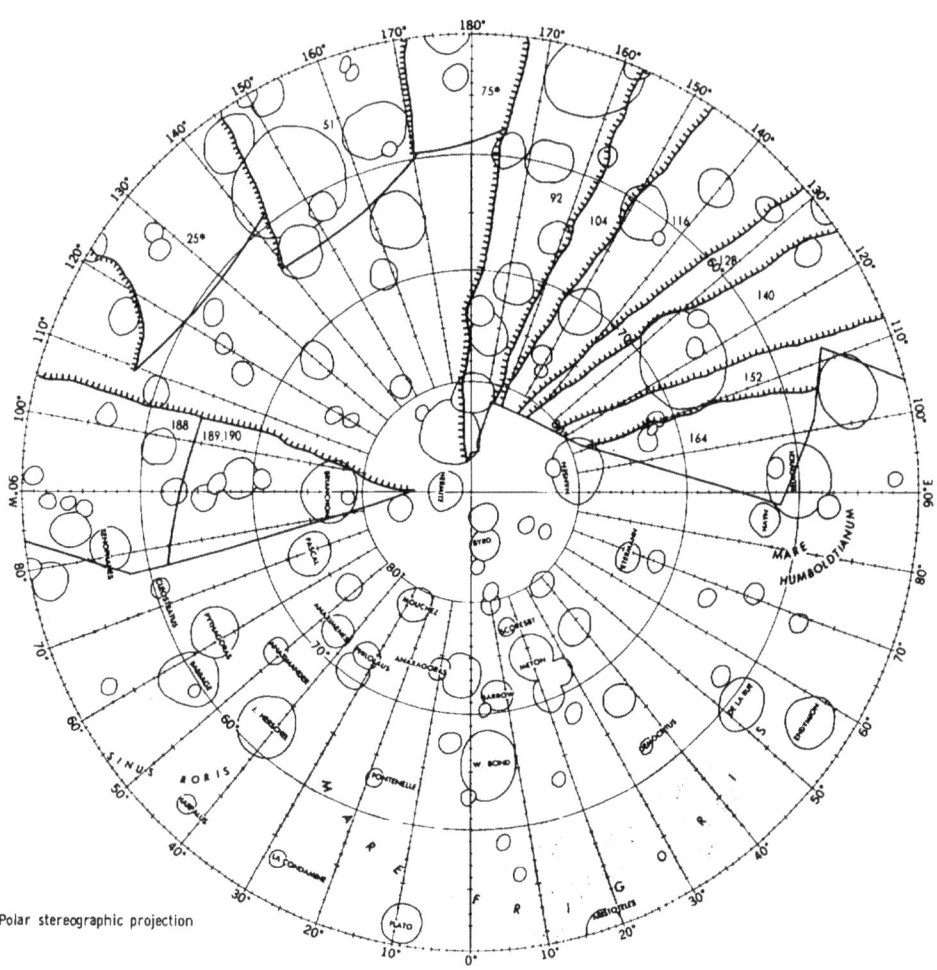

Polar stereographic projection

Terminator positions, for those
frames containing the terminator, are indicated by delineating
ticks. Numbers given are exposure numbers. An asterisk indicates the medium-resolution
frame is significantly degraded.

(c) North polar region.

FIGURE 11.—Photographic Indexes for selected mission IV medium-resolution frames of the far side.—Continued

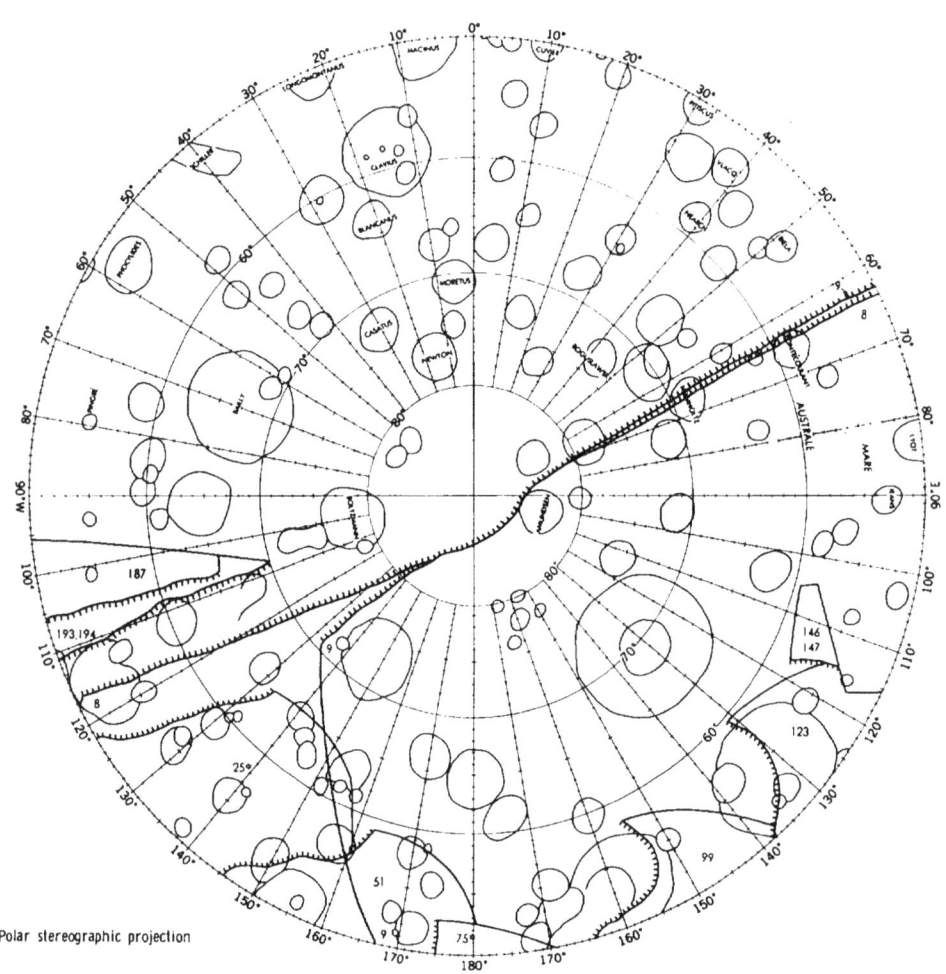

Polar stereographic projection

Terminator positions, for those
frames containing the terminator, are indicated by delineating
ticks. Numbers given are exposure numbers. An asterisk indicates the medium-resolution
frame is significantly degraded.

(d) *South polar region.*

FIGURE 11.—*Photographic Indexes for selected mission IV medium-resolution frames of the far side.—Concluded.*

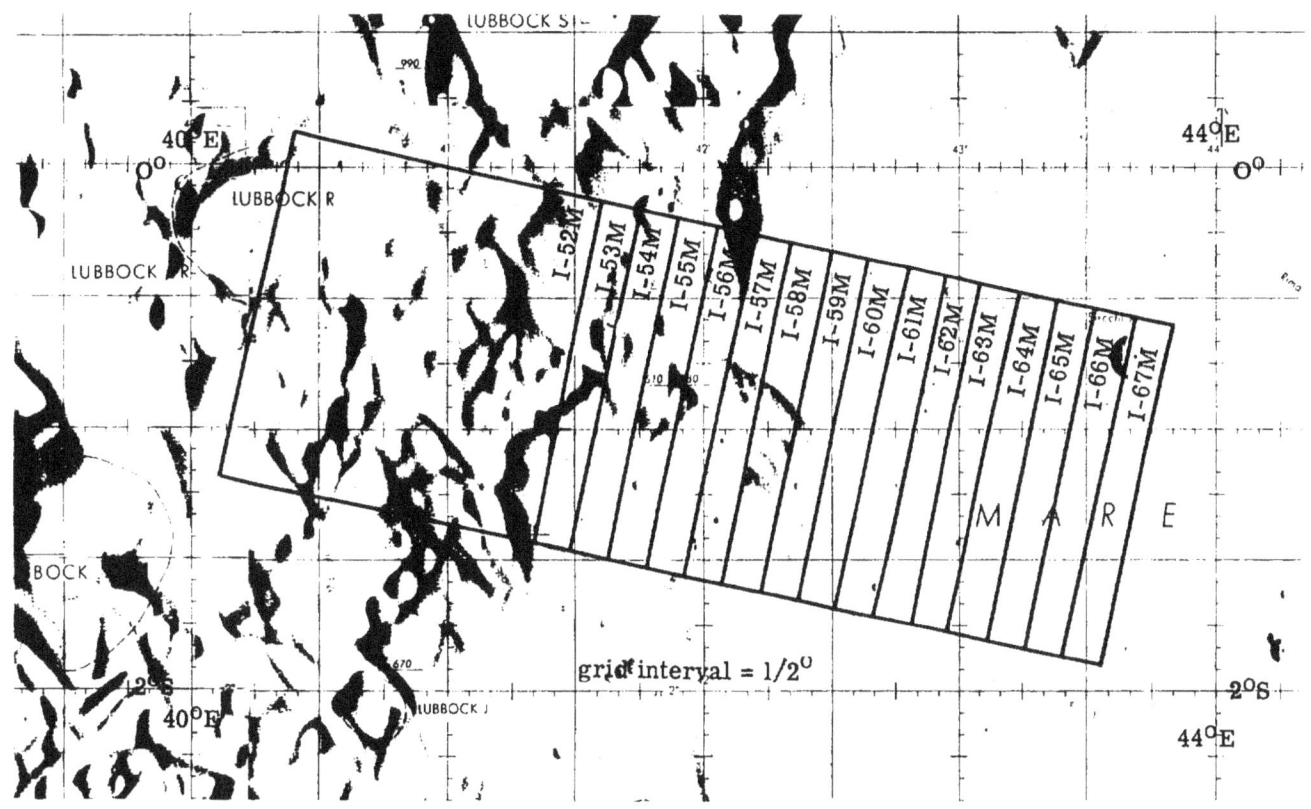

(a) Site IP-1.

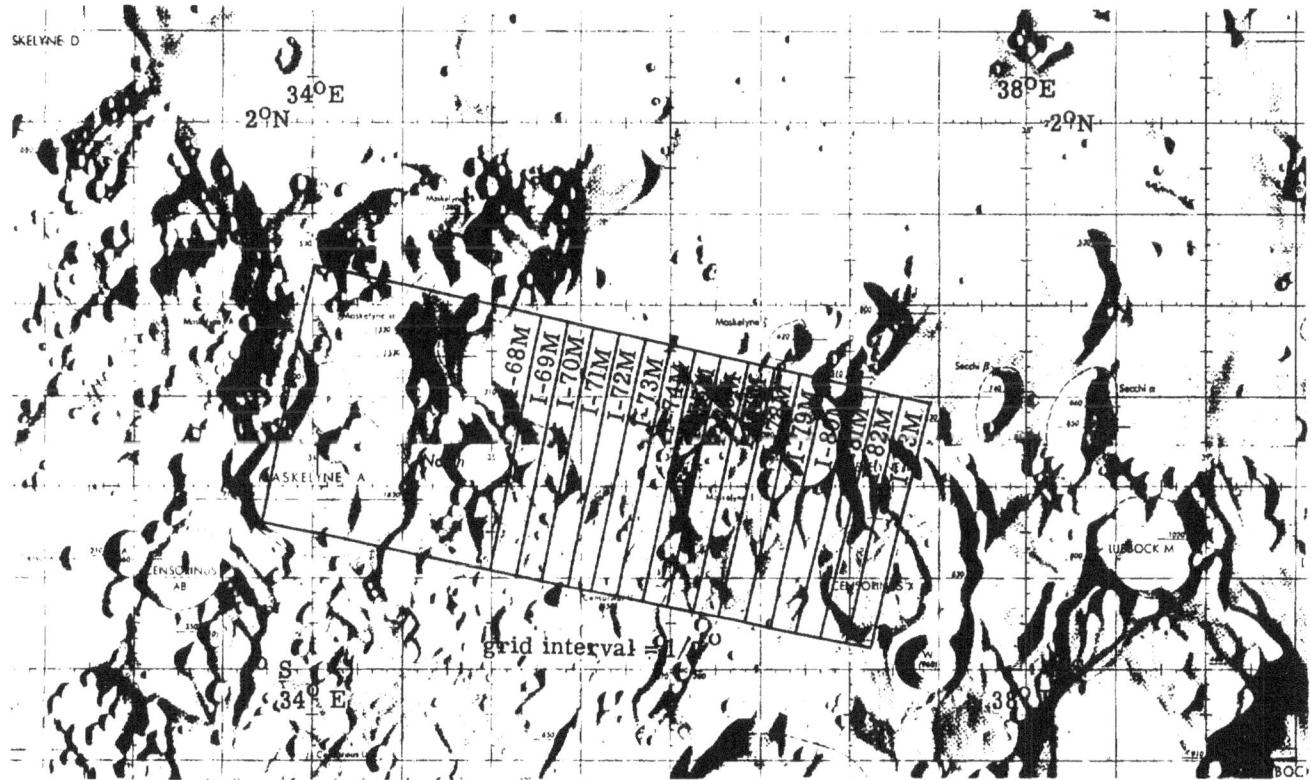

(b) Site IP-2.

FIGURE 12.—Photographic Indexes to mission I near-side sites.

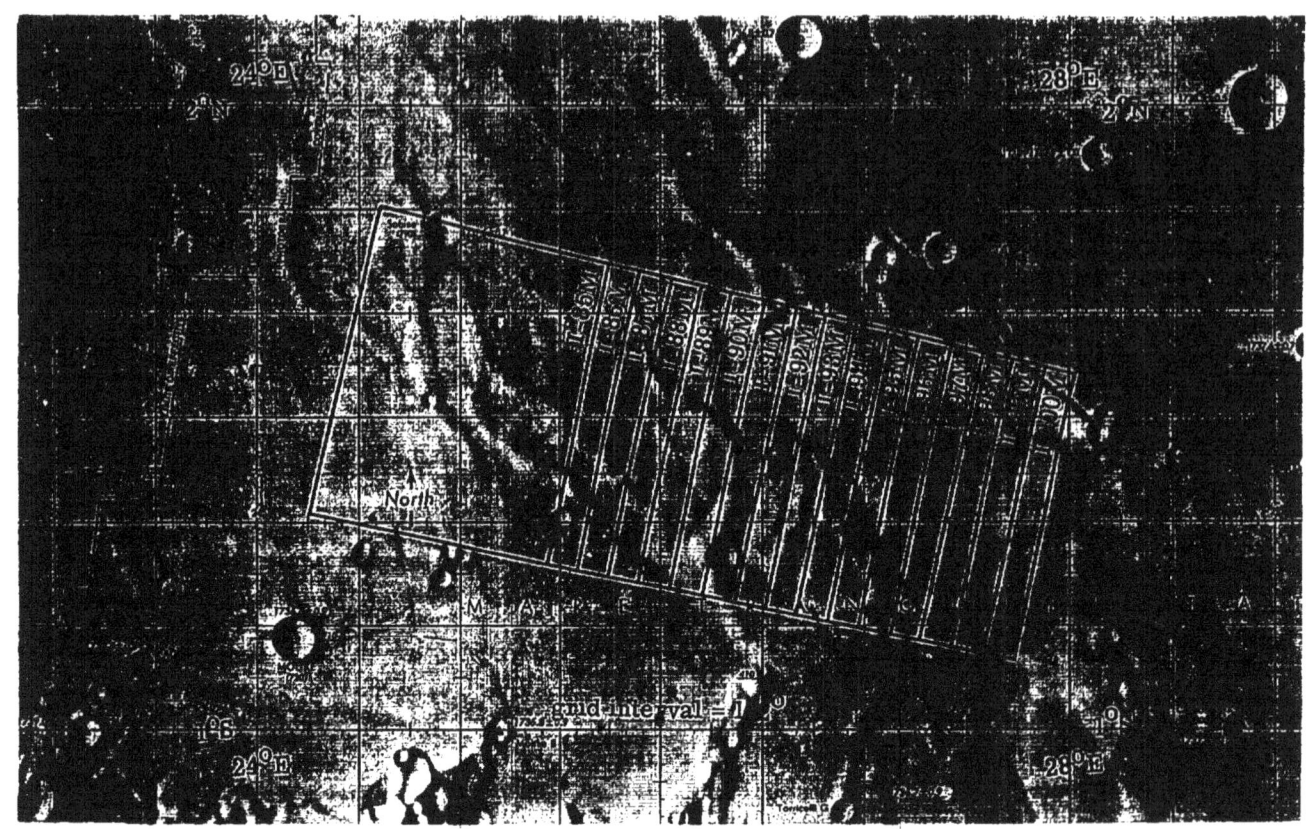

(c) Site IP-3.

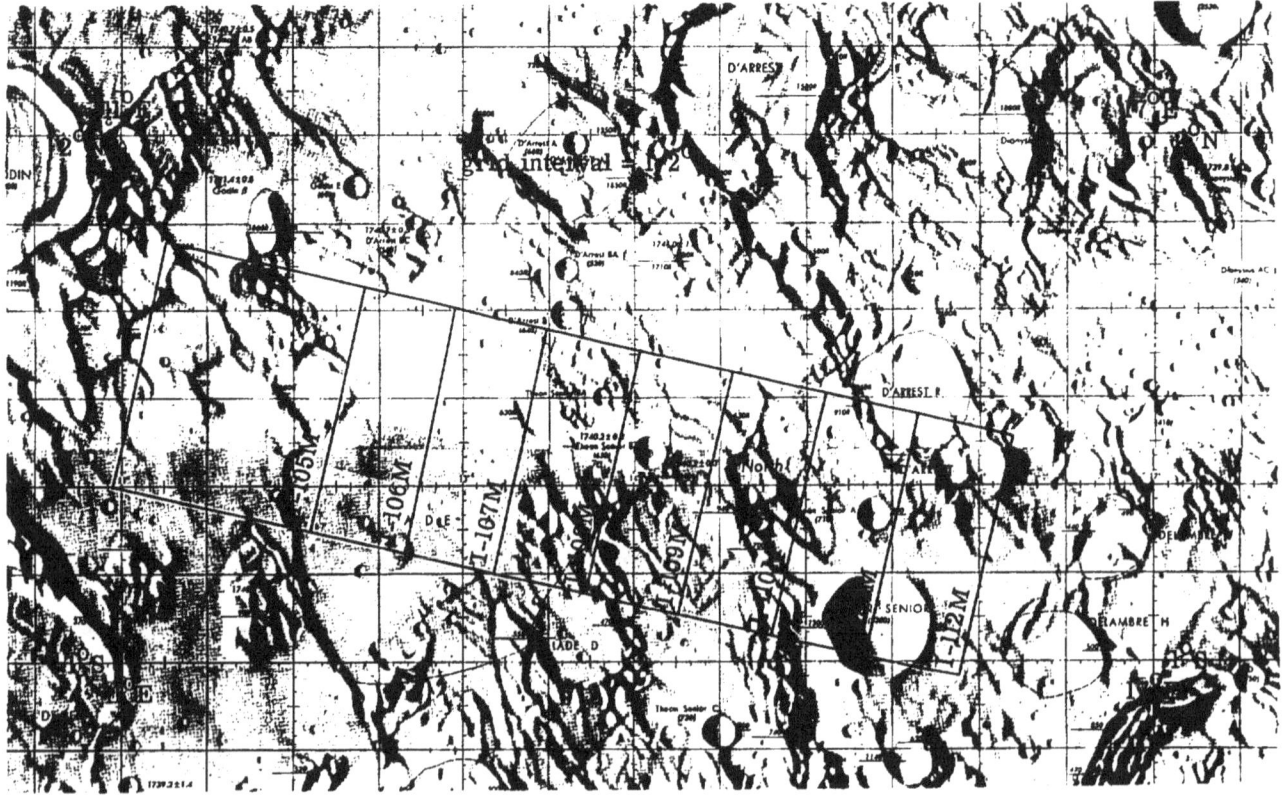

(d) Site IP-4.

FIGURE 12.—*Photographic Indexes to mission I near-side sites.*—Continued.

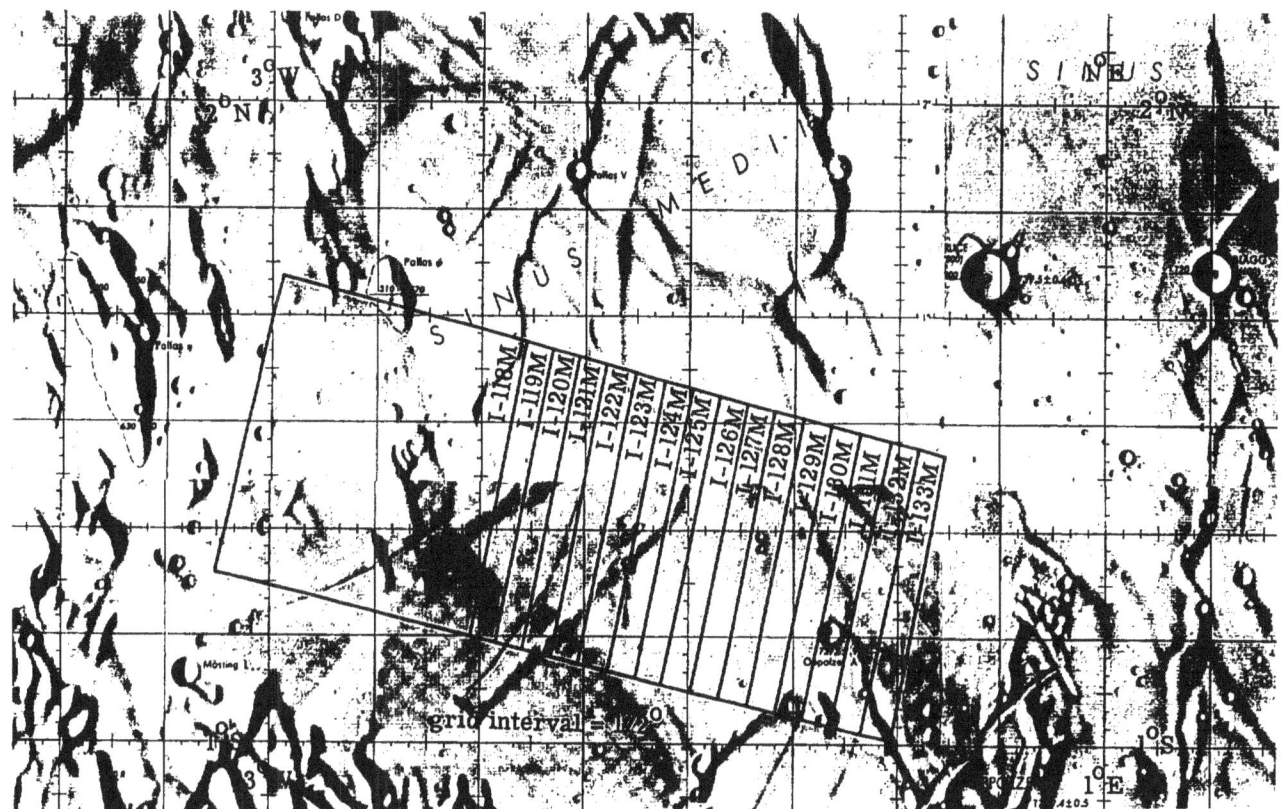

(e) Site IP-5.

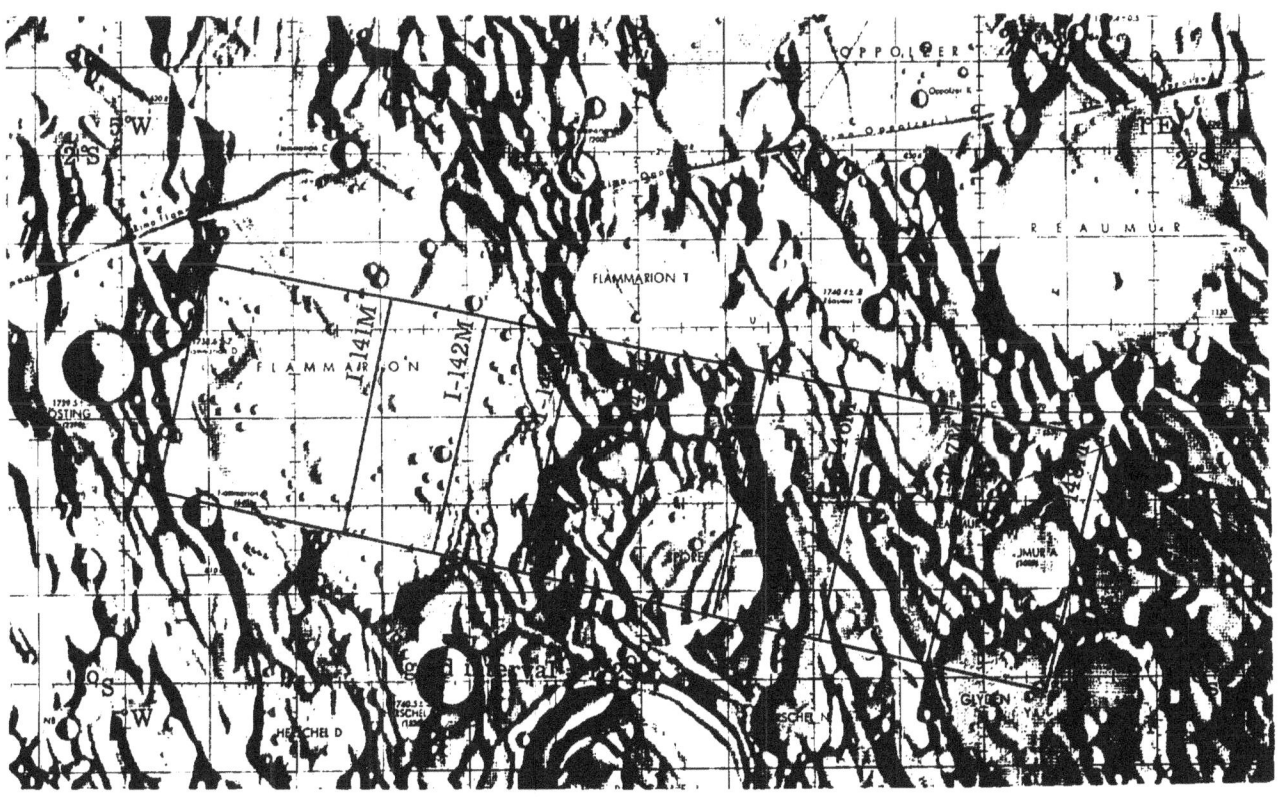

(f) Site IP-6.

FIGURE 12.—*Photographic Indexes to mission 1 near-side sites.*—Continued.

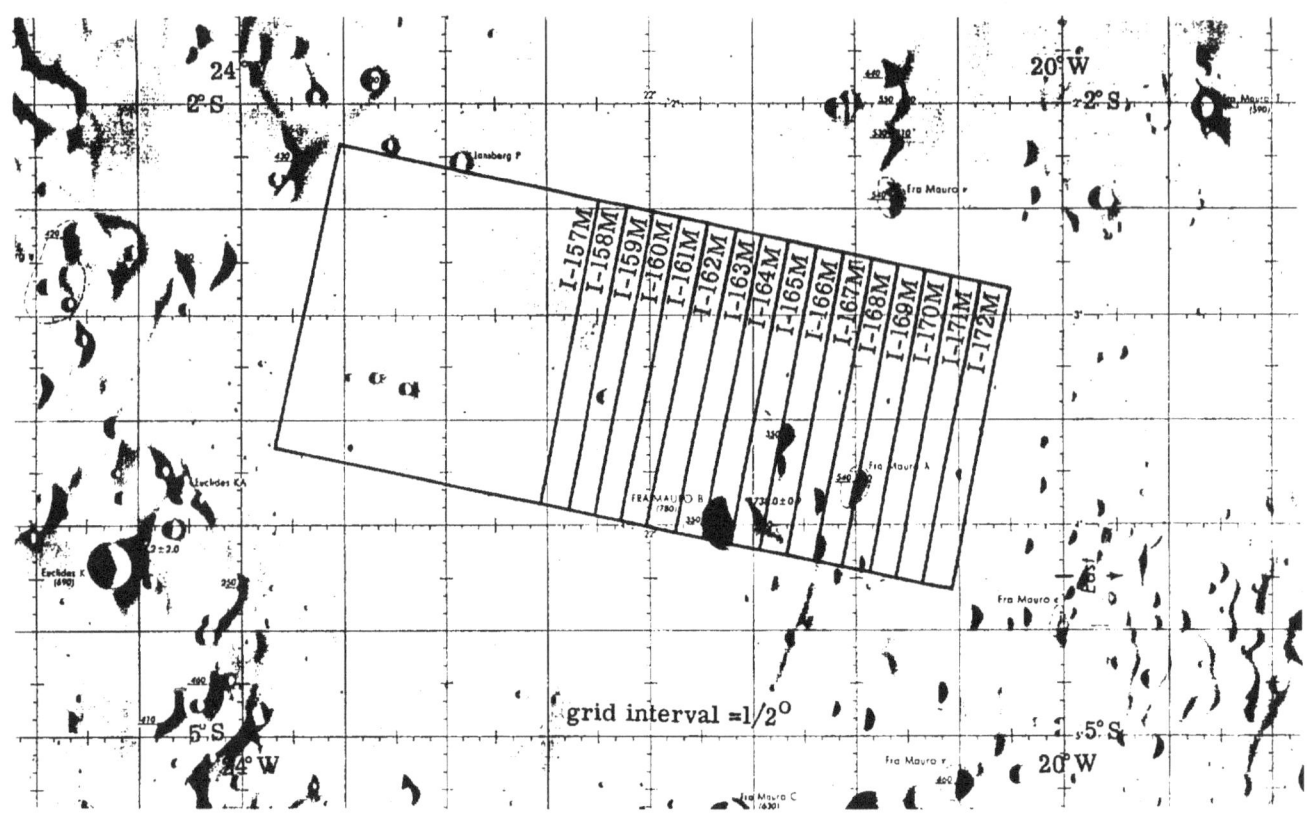

grid interval = 1/2°

(g) Site IP-7.

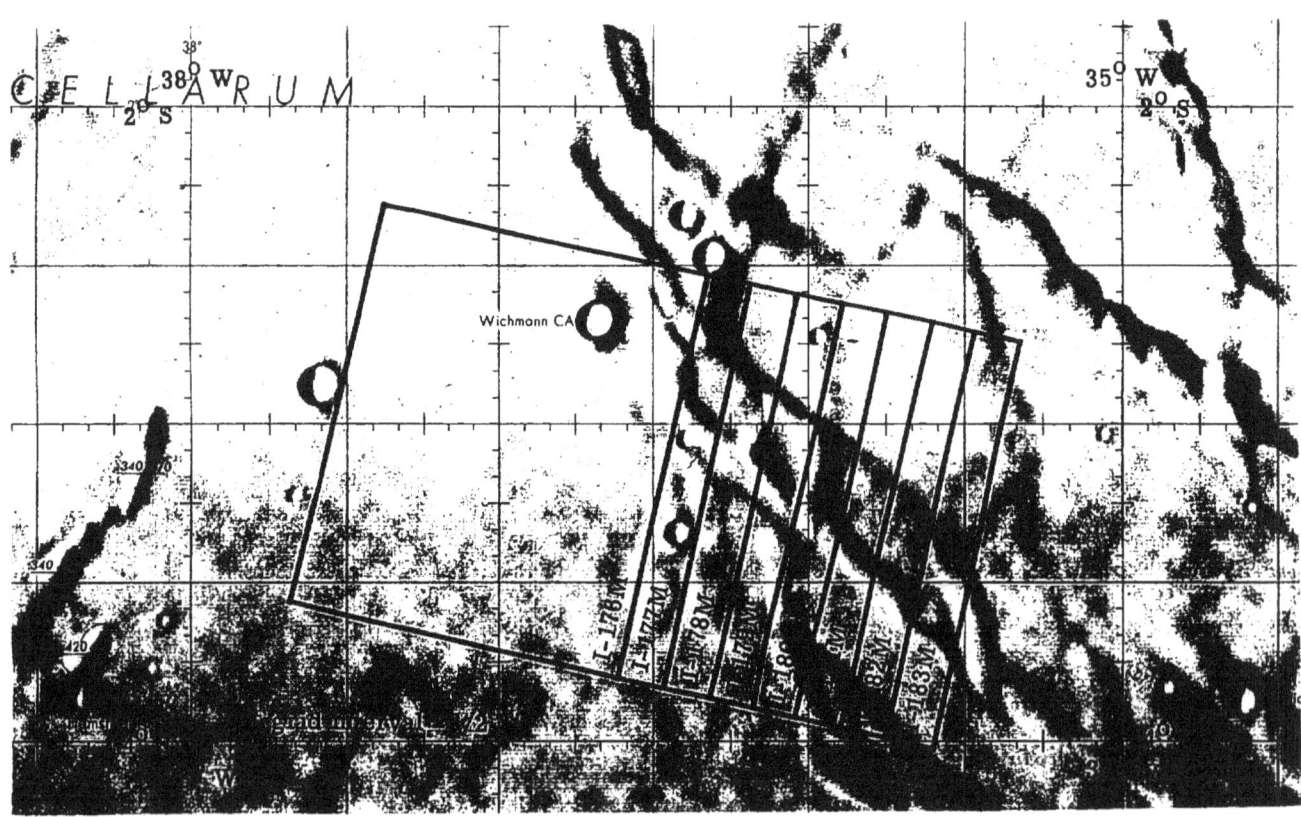

(h) Site IP-8.1.

FIGURE 12.—Photographic Indexes to mission I near-side sites.—Continued.

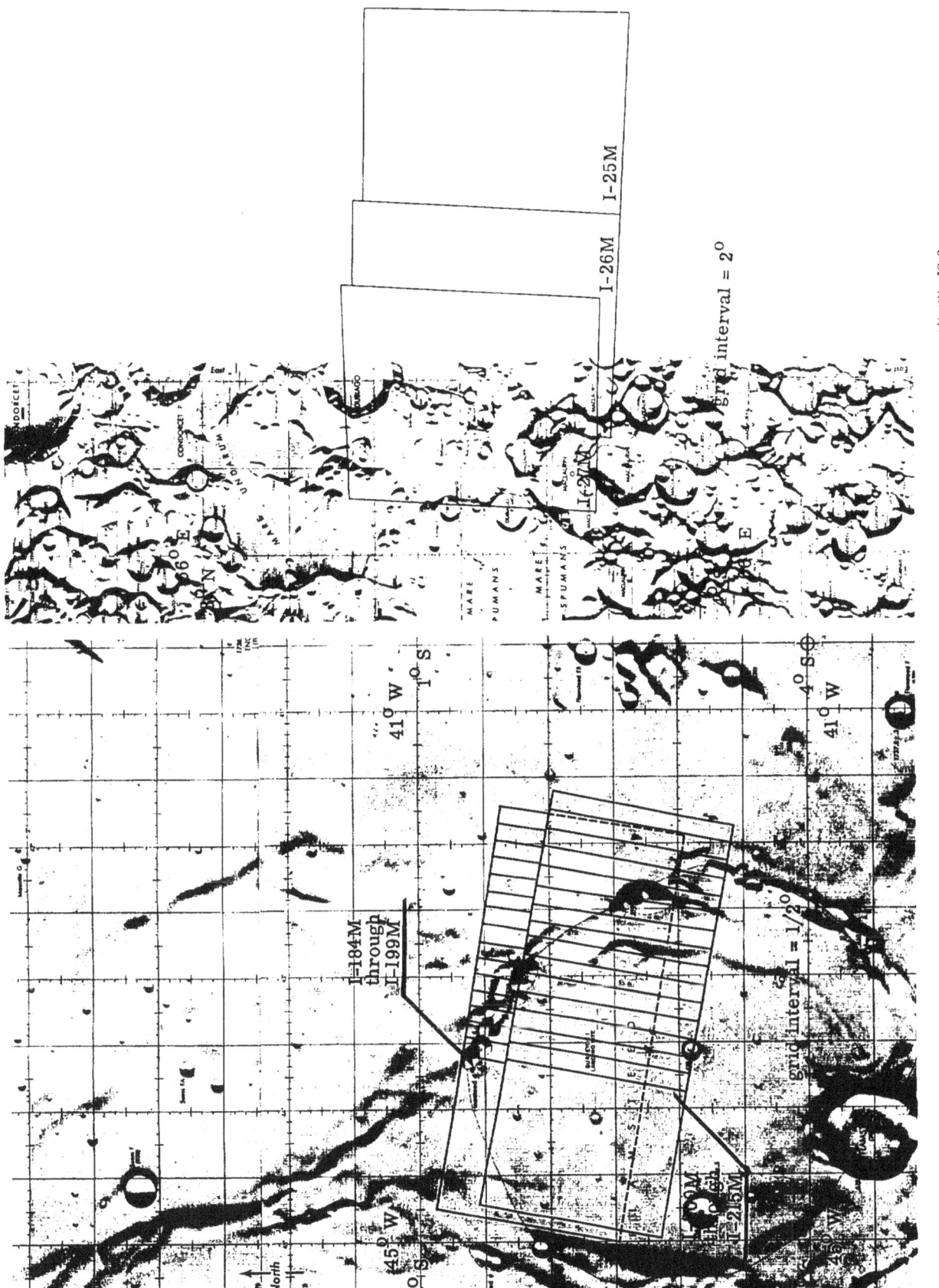

I-25M

I-26M

grid interval = 2°

(j) site IS-2.

I-184M through I-199M

grid interval = 1/2°

(i) Site IP-9.2.

FIGURE 12.—Photographic Indexes to mission 1 near-side sites.—Continued.

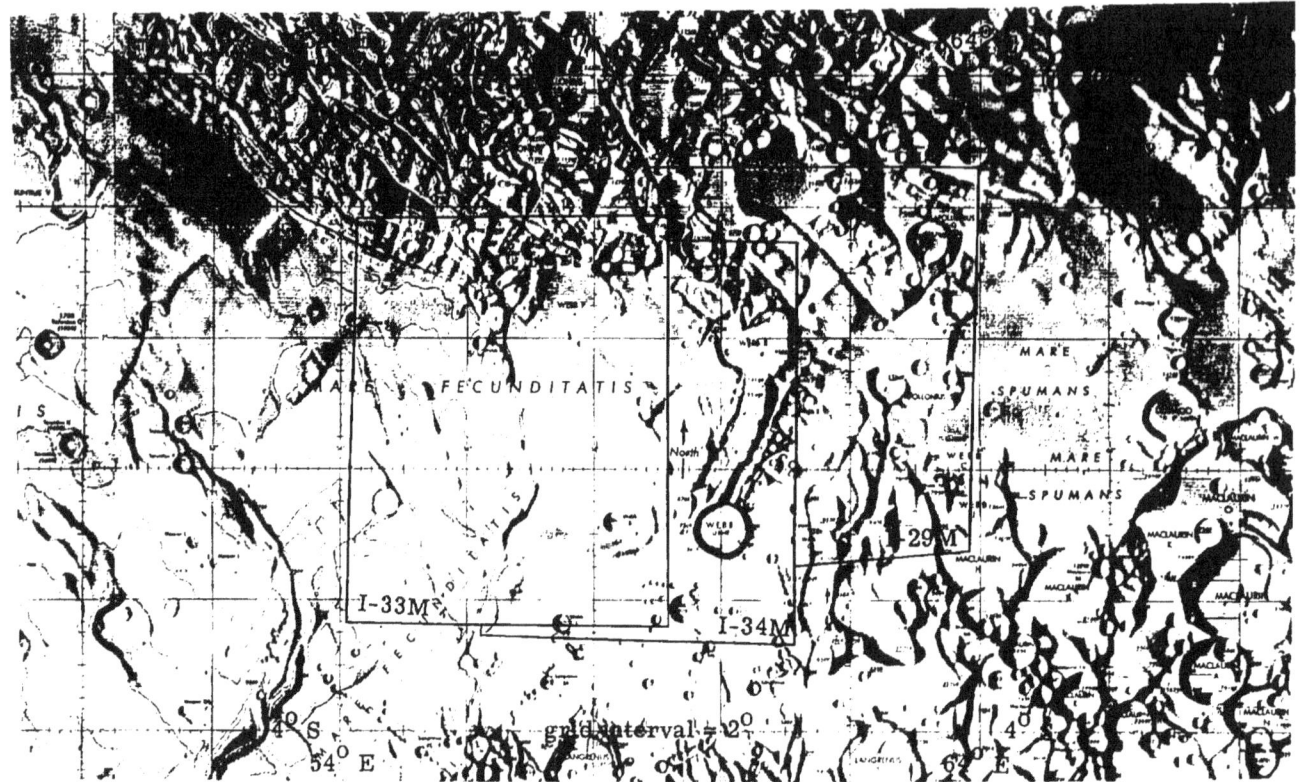

(k) Site IS-4.

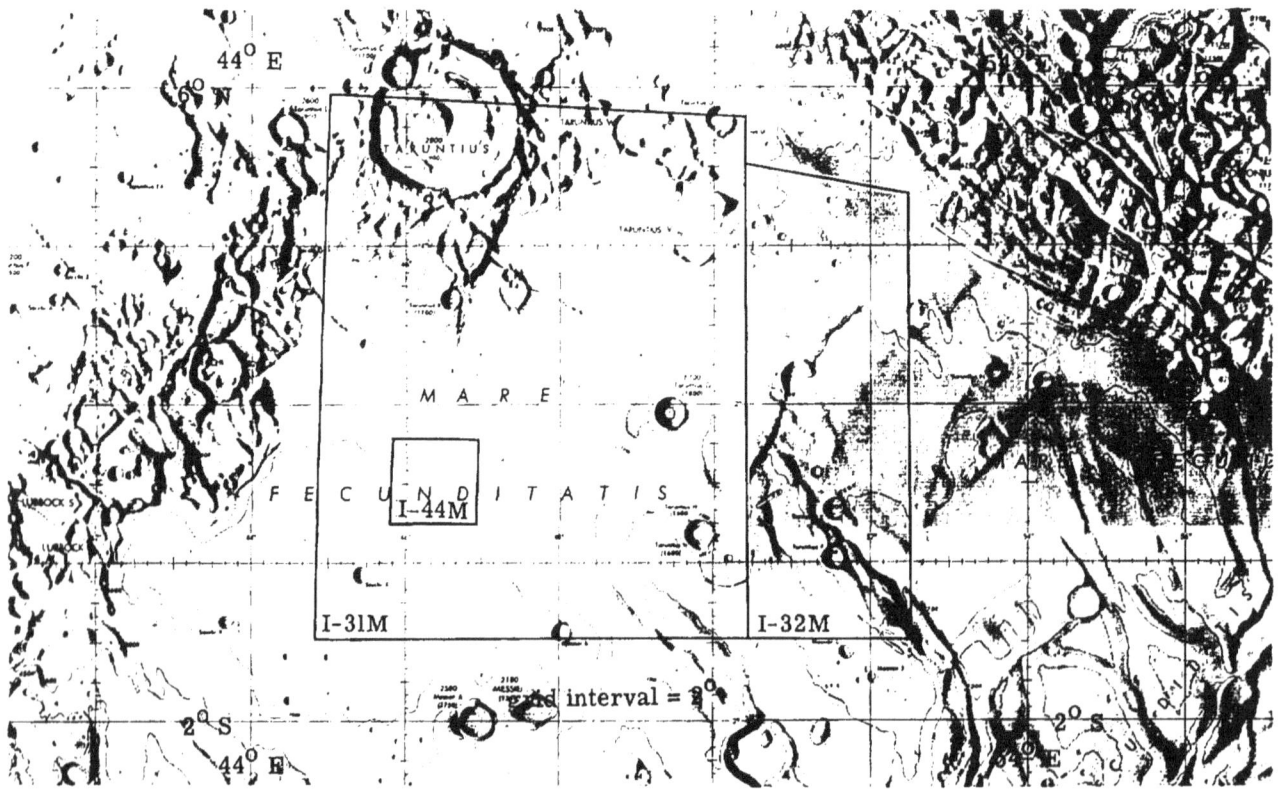

(l) Site IS-5.

FIGURE 12.—*Photographic Indexes to mission I near-side sites.*—Continued.

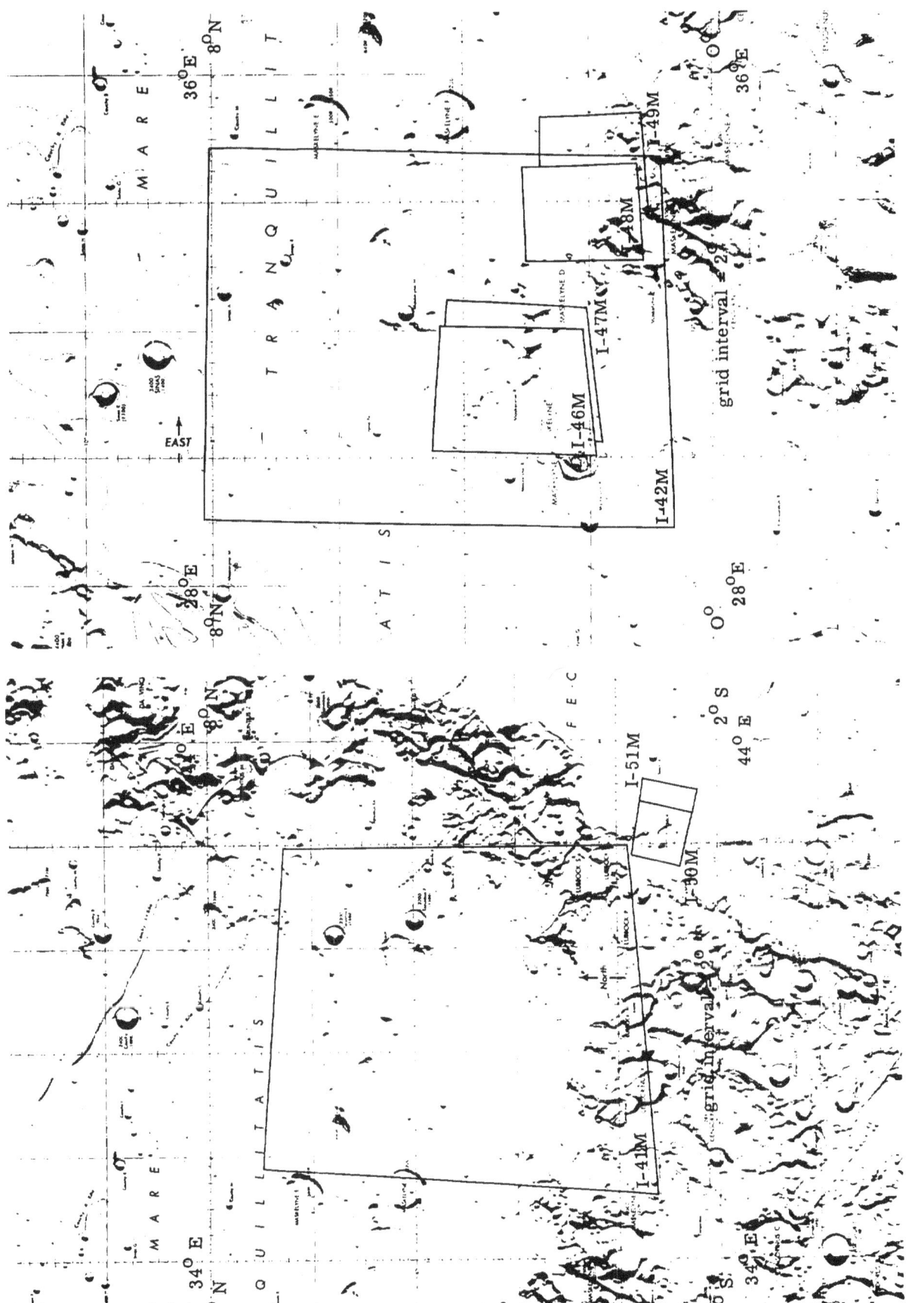

(n) Site IS-7.

(m) Site IS-6.

FIGURE 12.—Photographic indexes to mission I near-side sites.—Continued.

(p) *Site IS–10.*

(o) *Site IS–8.*

FIGURE 12.—*Photographic Indexes to mission I near-side sites.*—Continued.

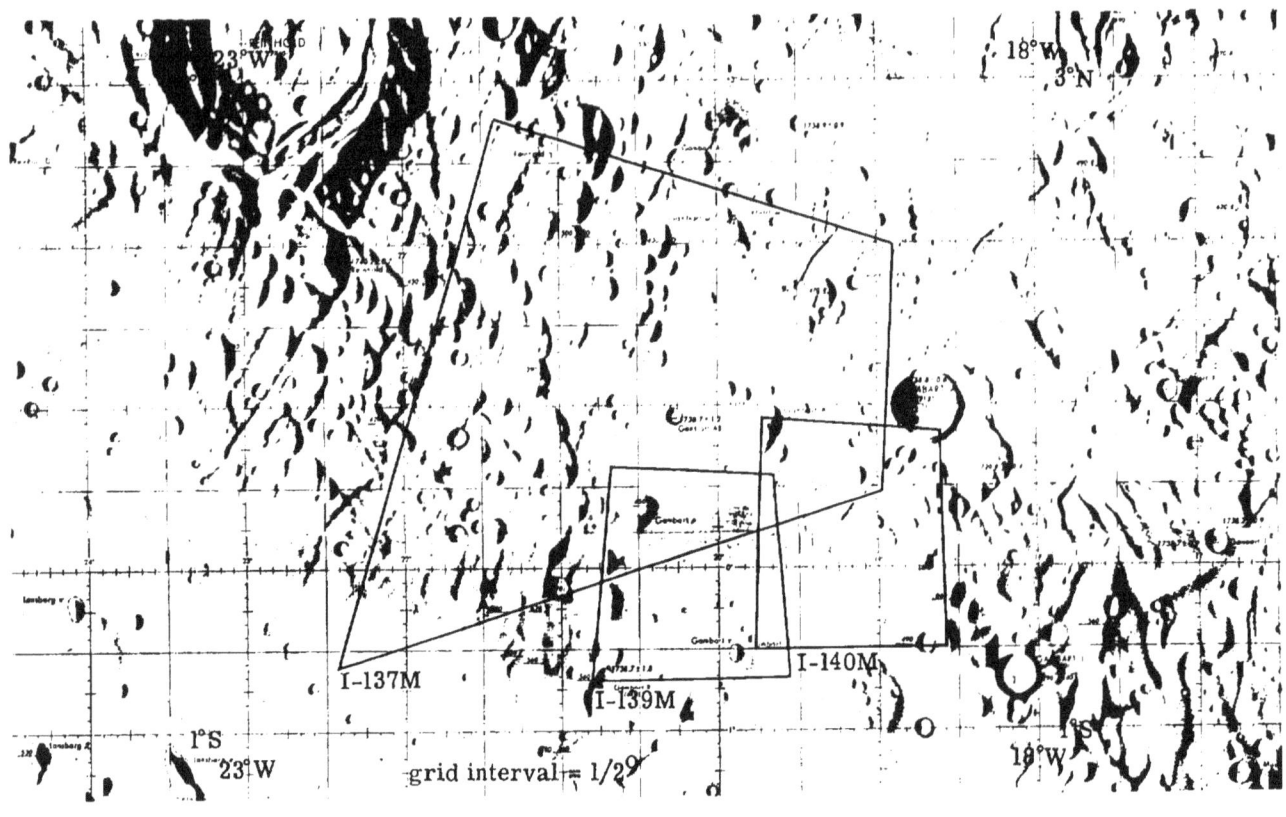

(s) Site IS-14.

(t) Site IS-15.

FIGURE 12.—*Photographic Indexes to mission I near-side sites.*—Continued.

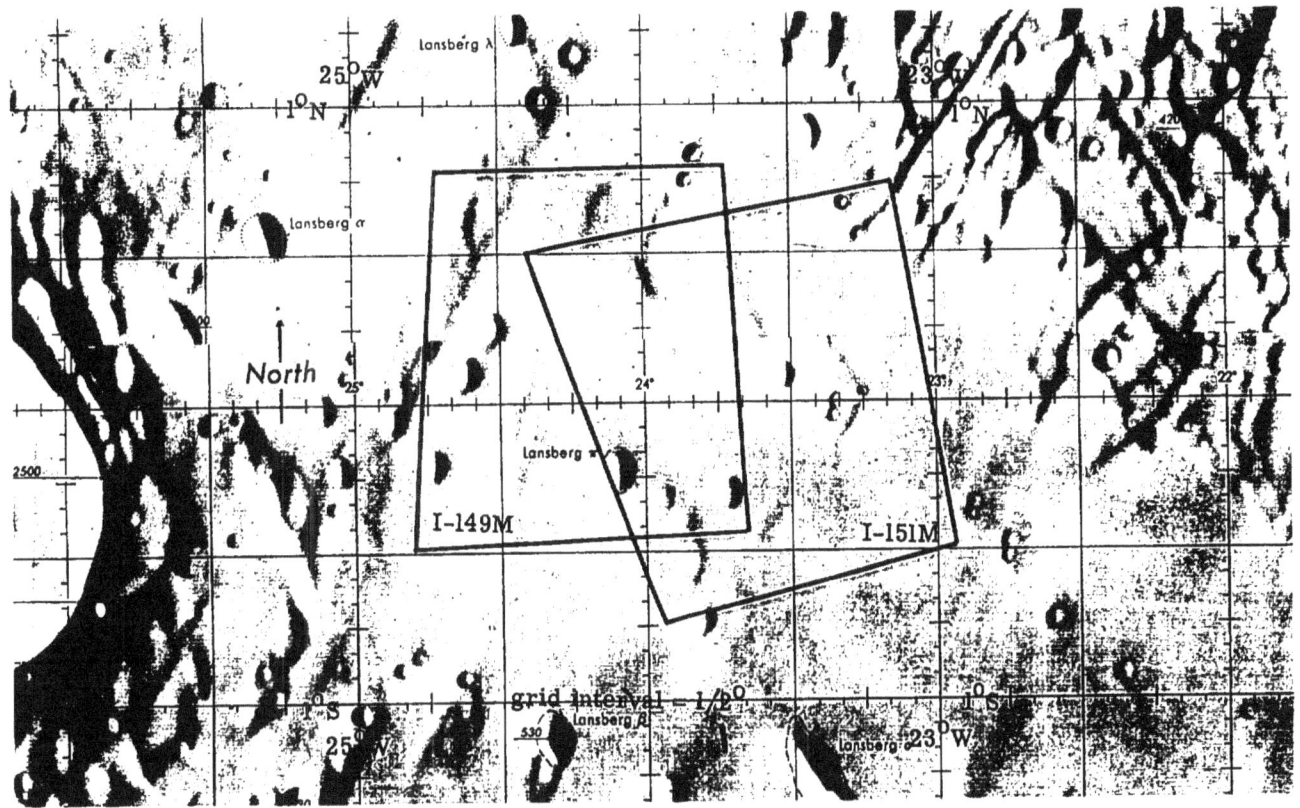

(u) Site IS-16.

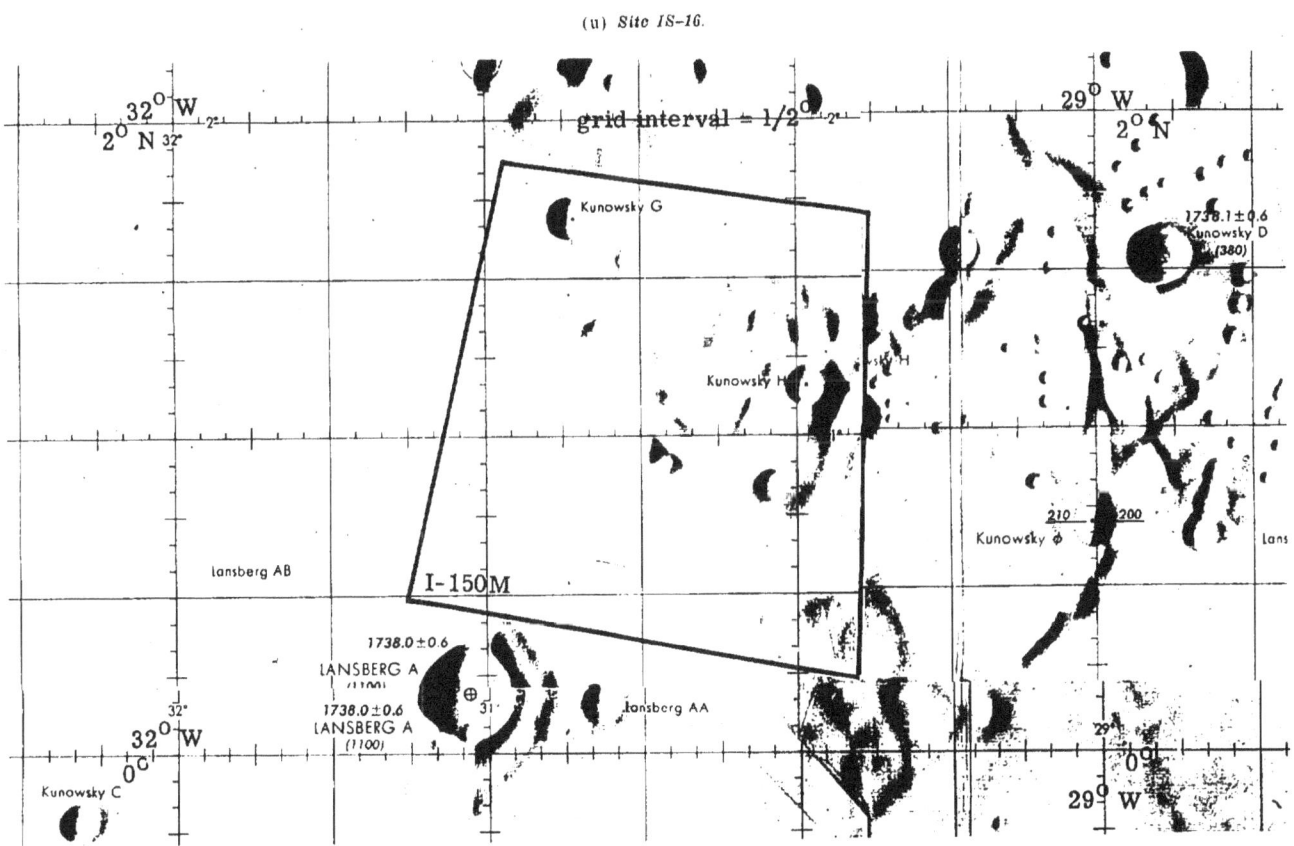

(v) Site IS-17.

FIGURE 12.—Photographic Indexes to mission I near-side sites.—Continued.

57

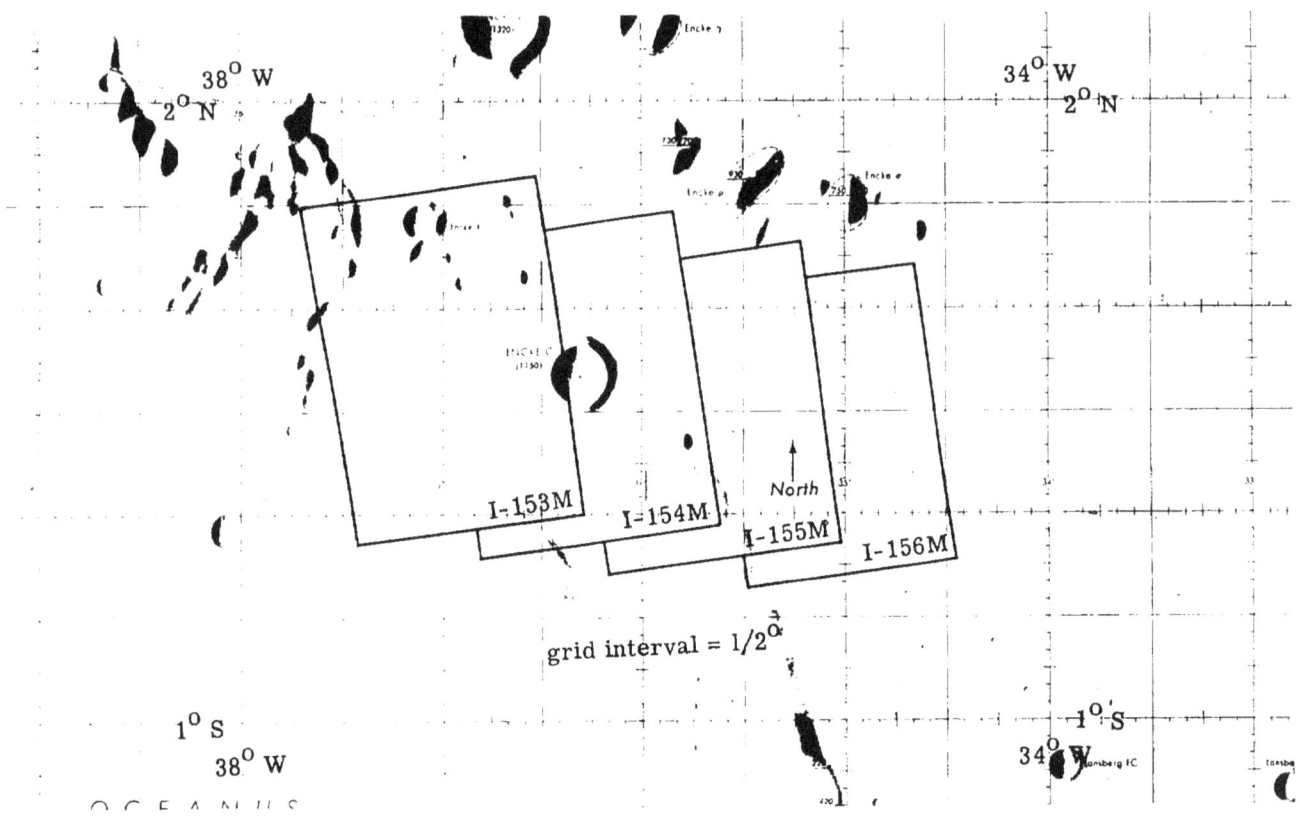

38° W
2° N

34° W
2° N

I-153M   I-154M

North

I-155M

I-156M

grid interval = 1/2°

1° S

1° S

38° W

34° W

O C E A N U S

(w) Site 18–18.

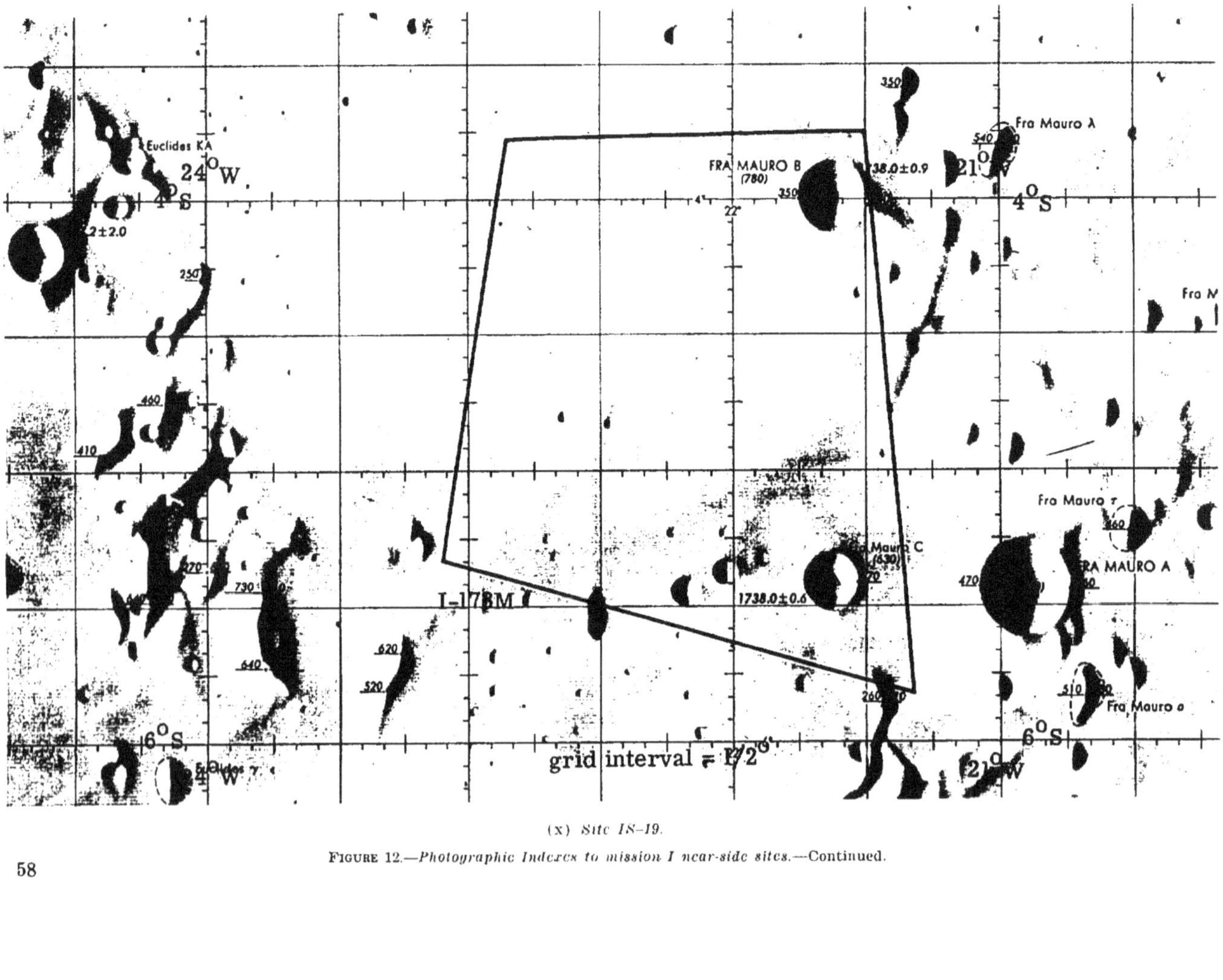

Euclides KA

24° W
4° S

FRA MAURO B
(780)

1738.0±0.9

Fra Mauro λ

21°

4° S

Fra M

Fra Mauro γ

Mauro C
(630)

FRA MAURO A

I-173M

1738.0±0.6

Fra Mauro α

6° S

grid interval = 1/2°

6° S

21°

(x) Site 18–19.

FIGURE 12.—Photographic Indexes to mission I near-side sites.—Continued.

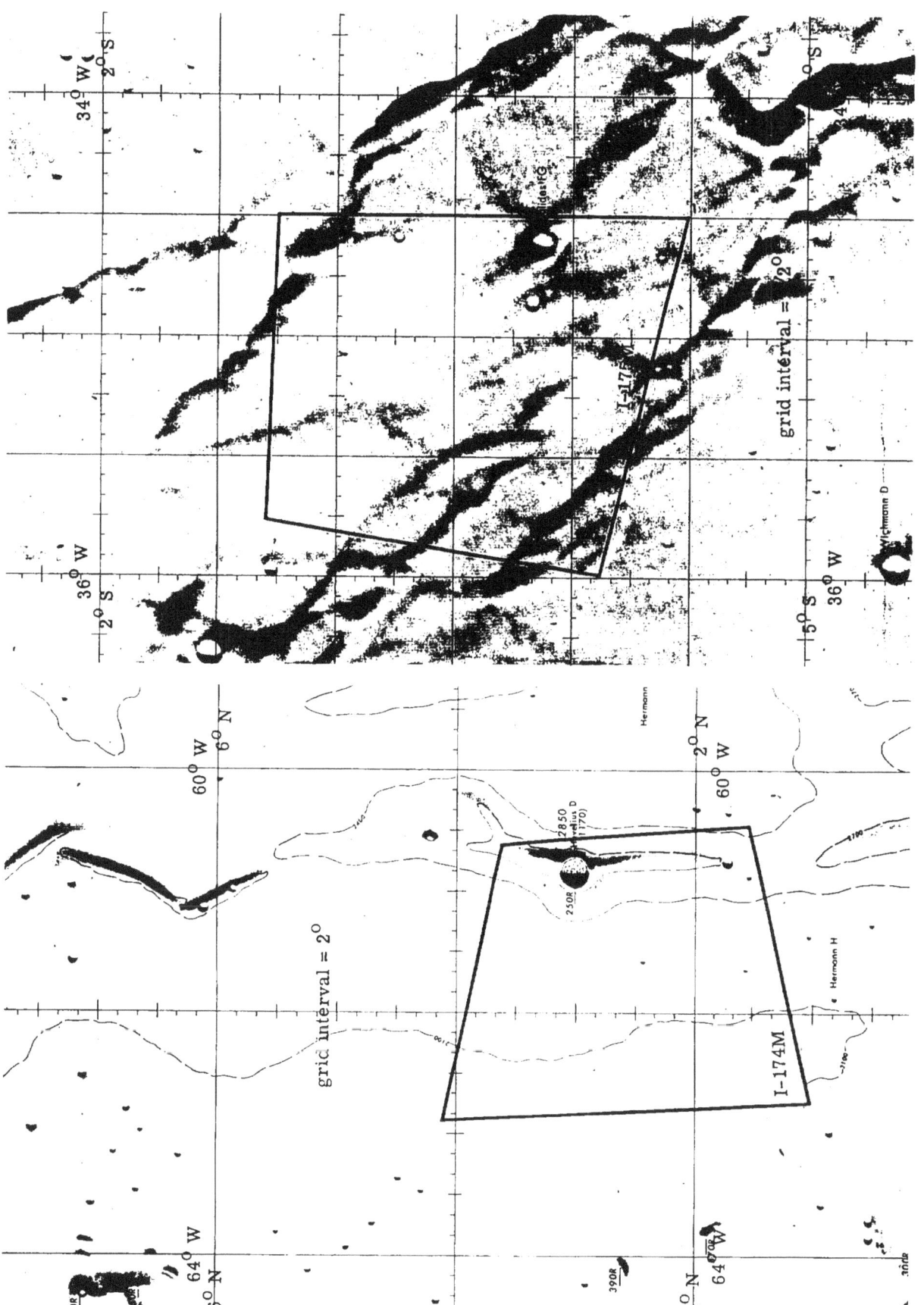

grid interval = 2°

(z) Site IS-21.

grid interval = 2°

(y) Site IS-20.

Figure 12.—Photographic Indexes to mission 1 near-side sites.—Concluded.

59

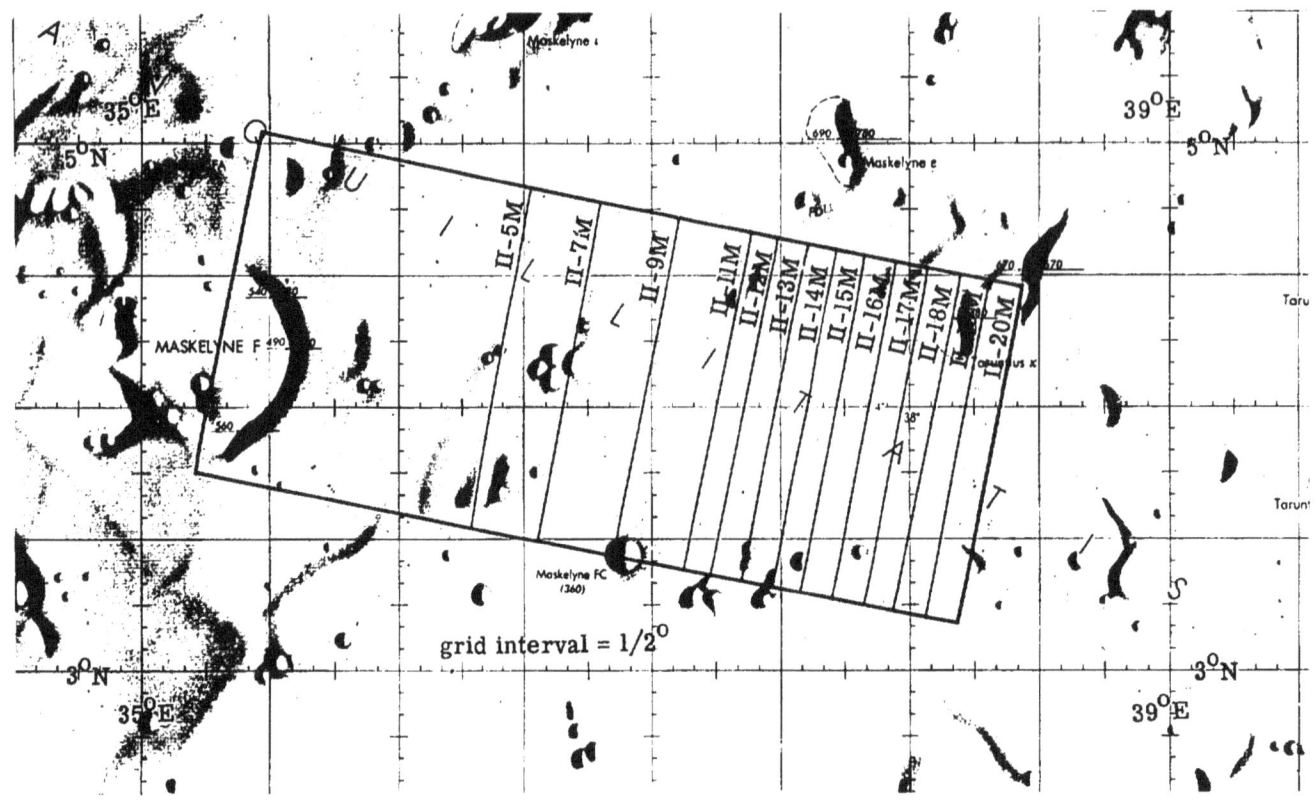

(a) IIP-1.

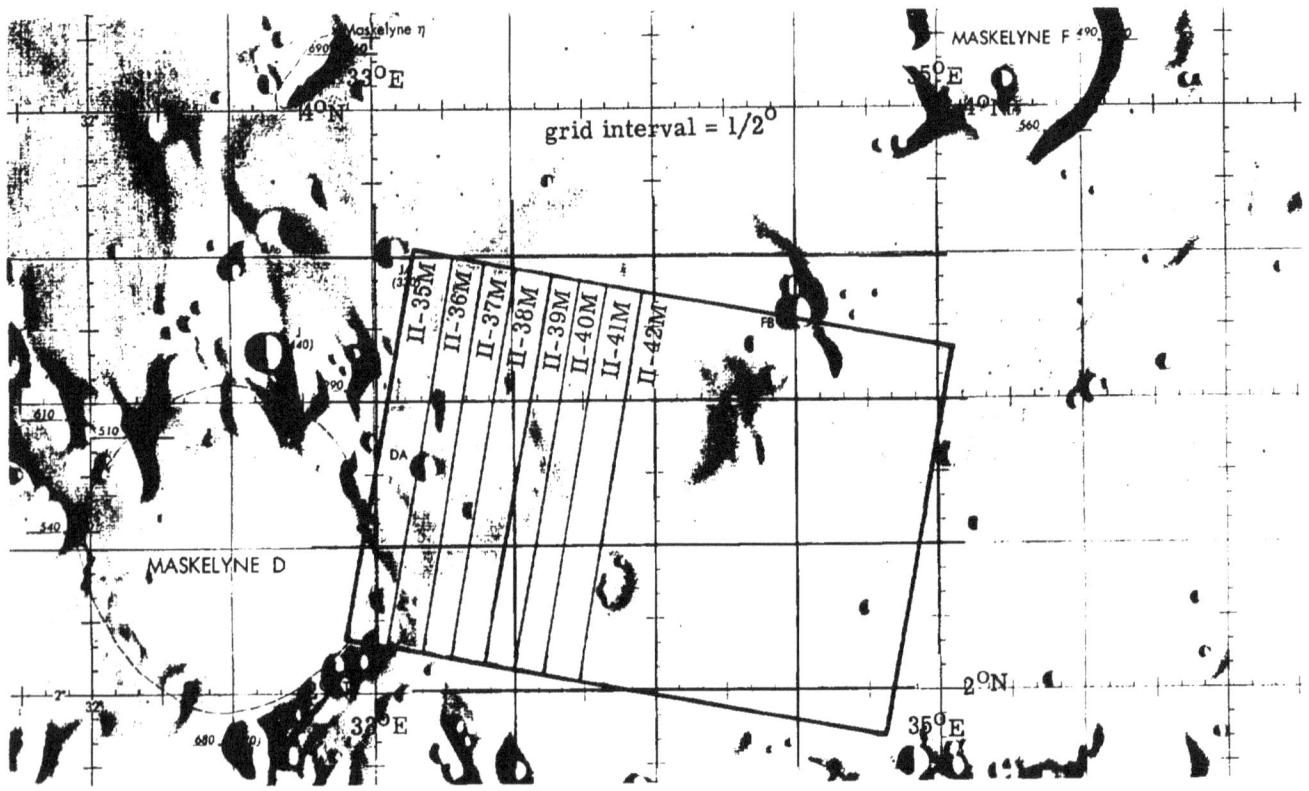

(b) Site IIP-2.

FIGURE 13.—Photographic Indexes to mission II near-side sites.

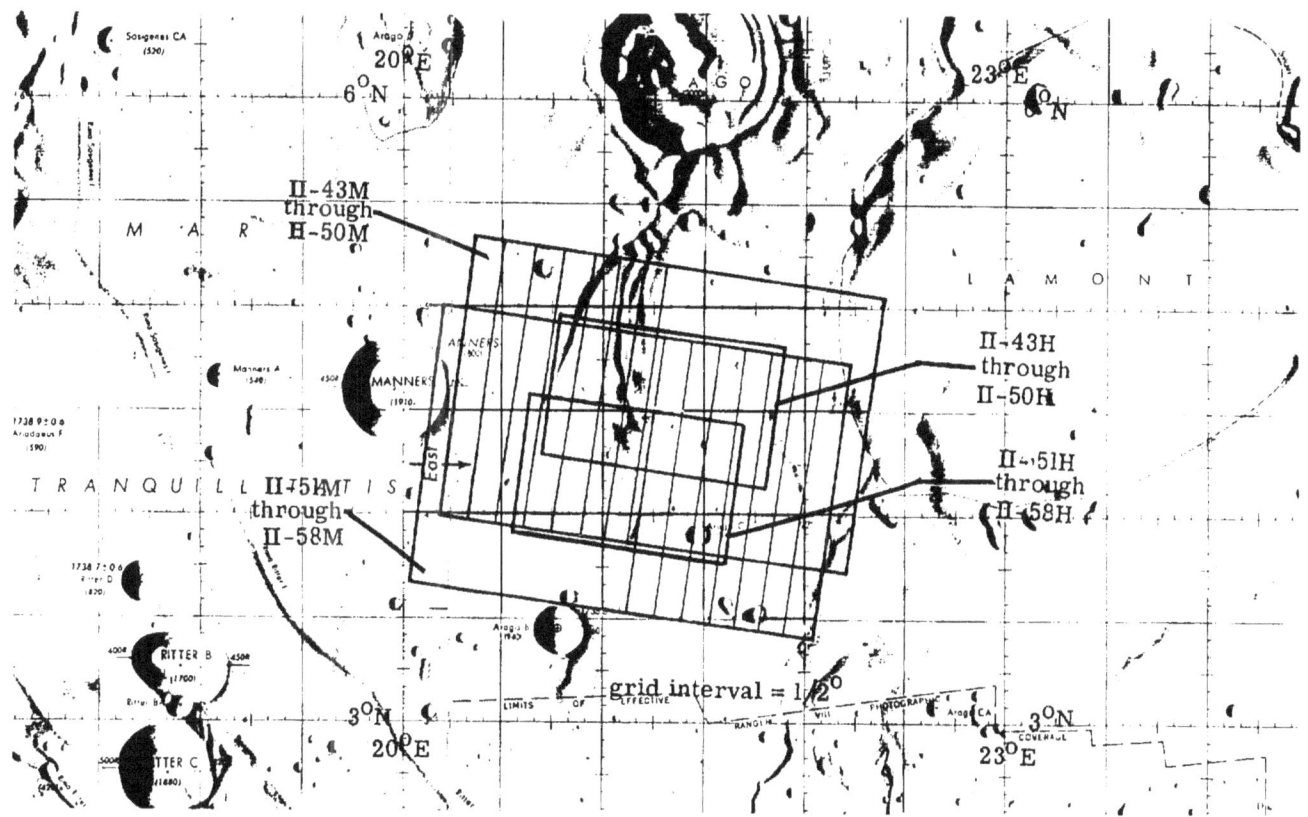

(c) *Site IIP-3.*

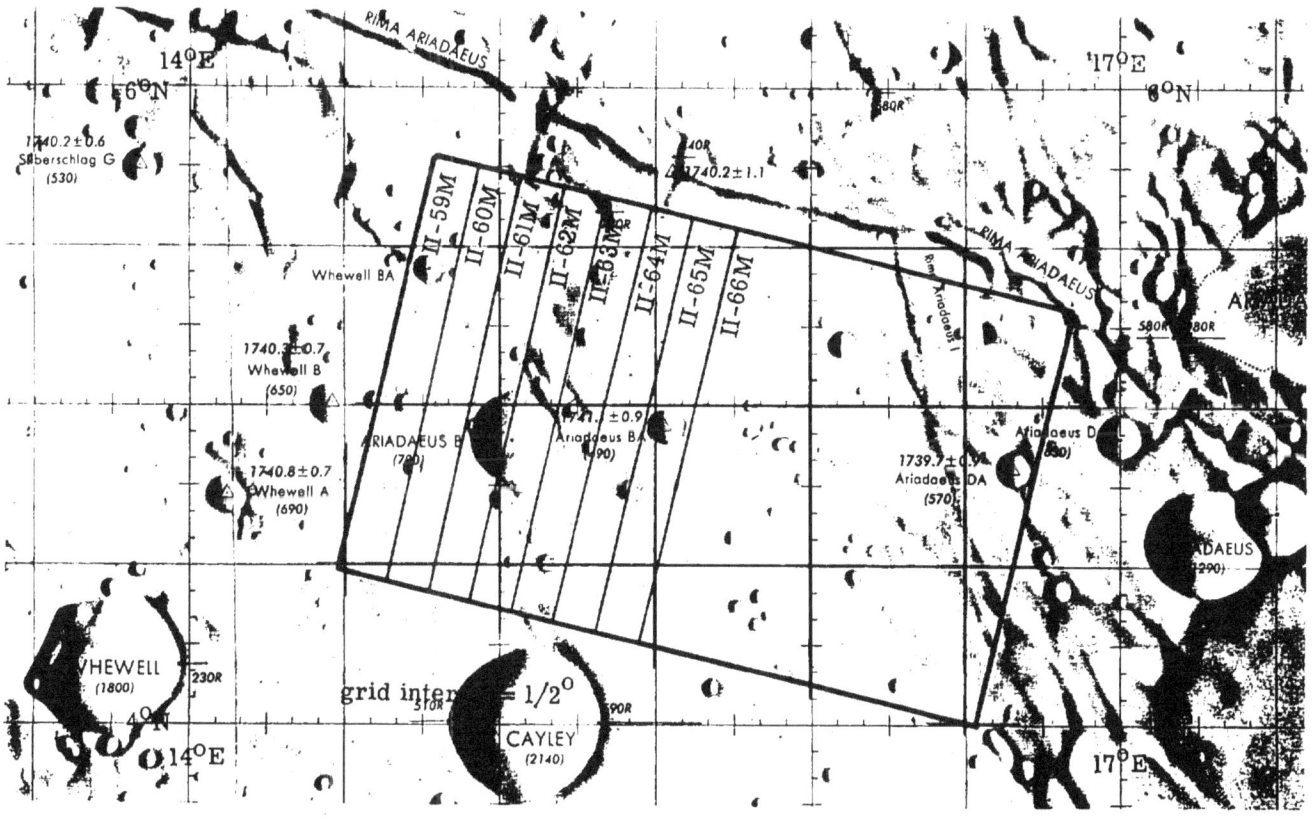

(d) *Site IIP-4.*

FIGURE 13.—*Photographic Indexes to mission II near-side sites.*—Continued.

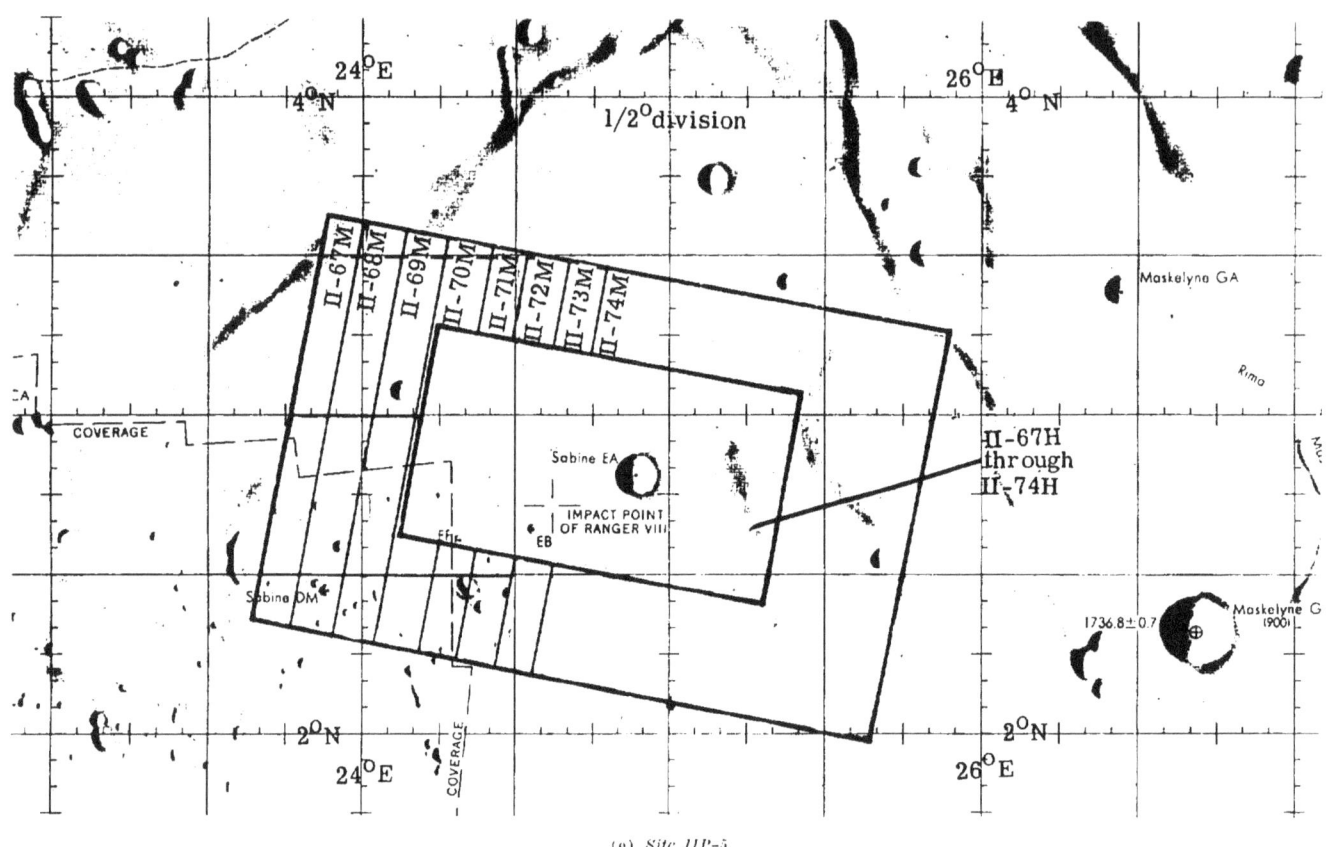

(e) Site HP-5.

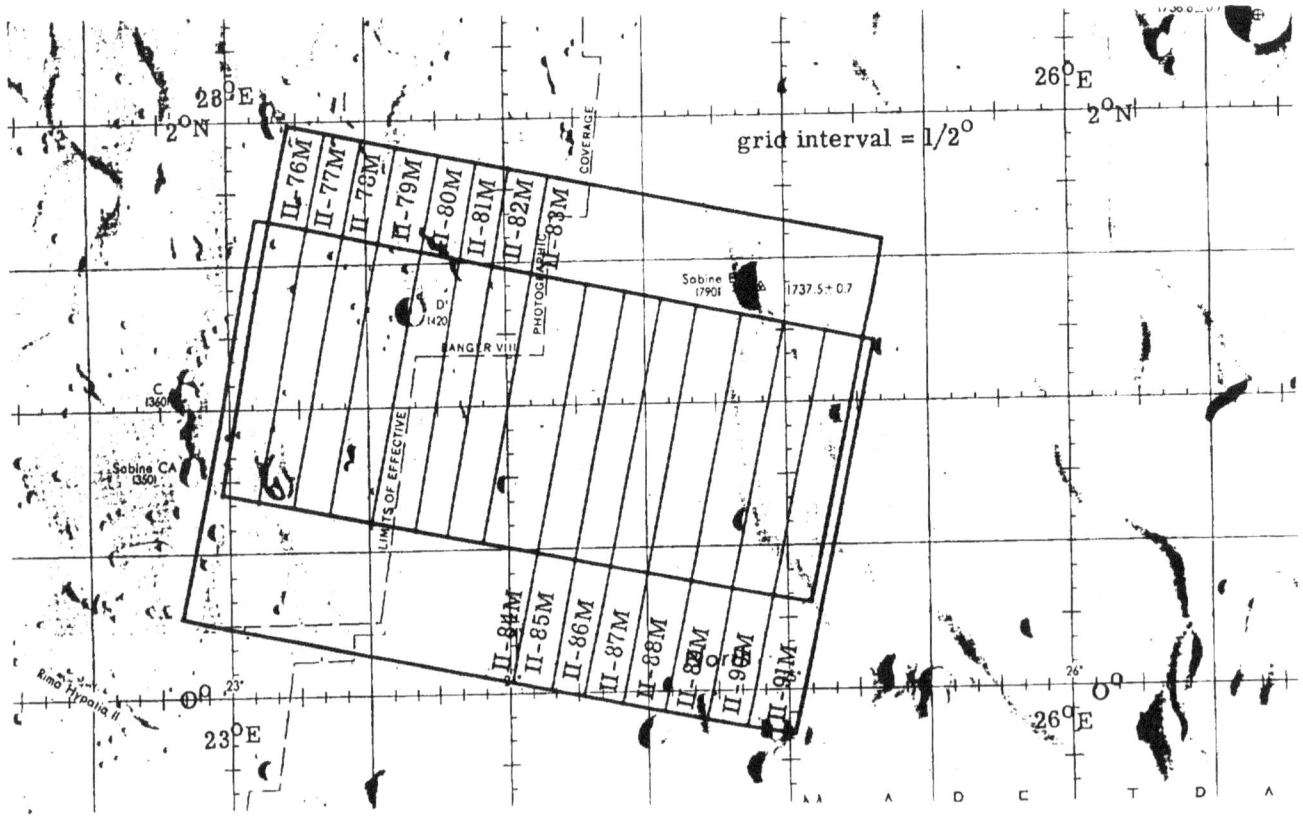

(f) Site HP-6.

FIGURE 13.—Photographic Indexes to mission II near-side sites.—Continued.

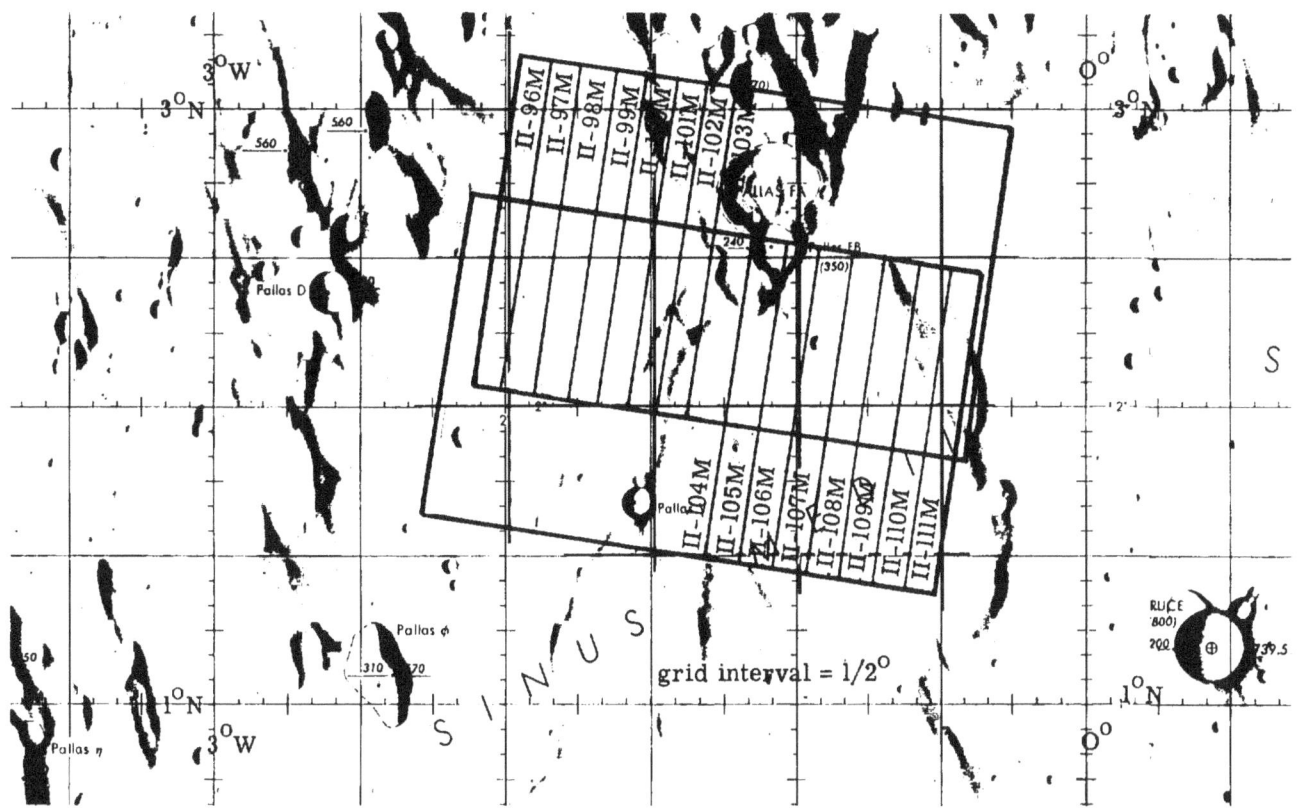

(g) Site IIP-7.

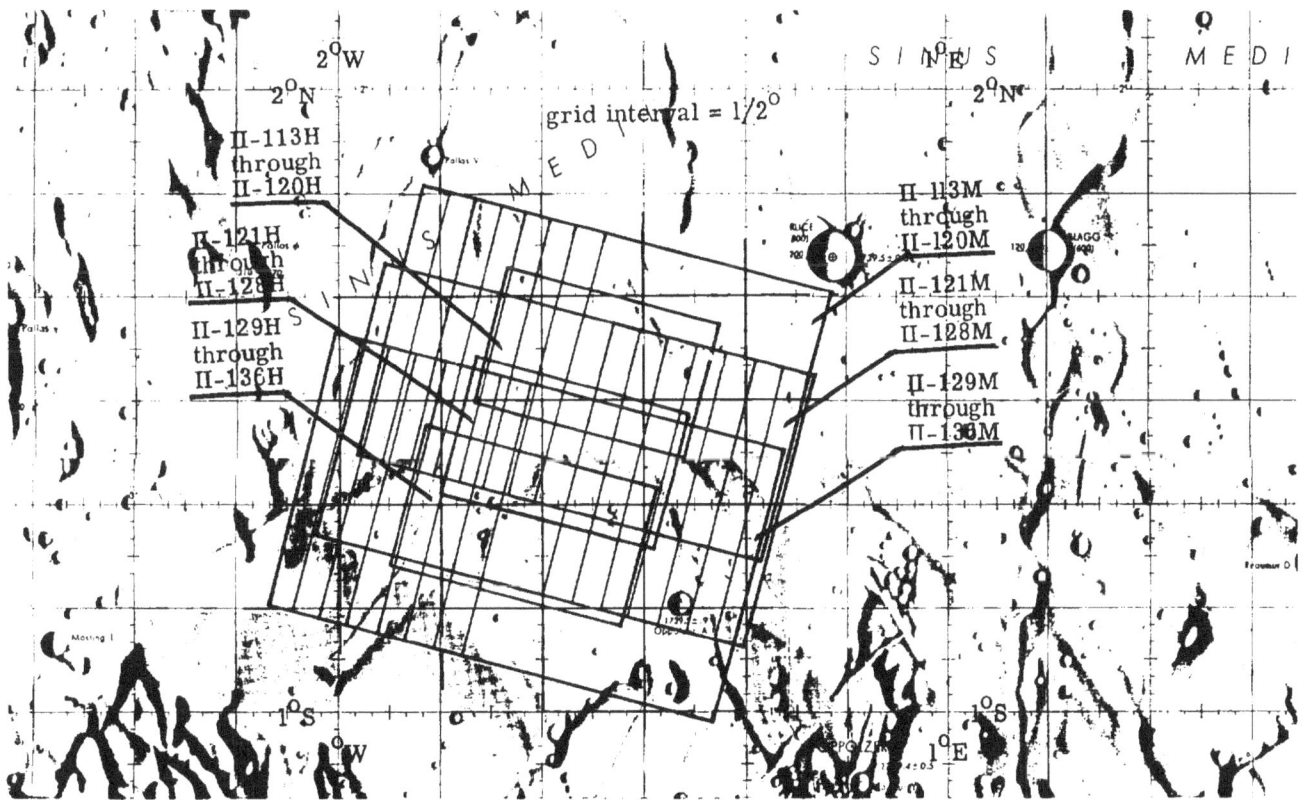

(h) Site IIP-8.

FIGURE 13.—*Photographic Indexes to mission II near-side sites.*—Continued.

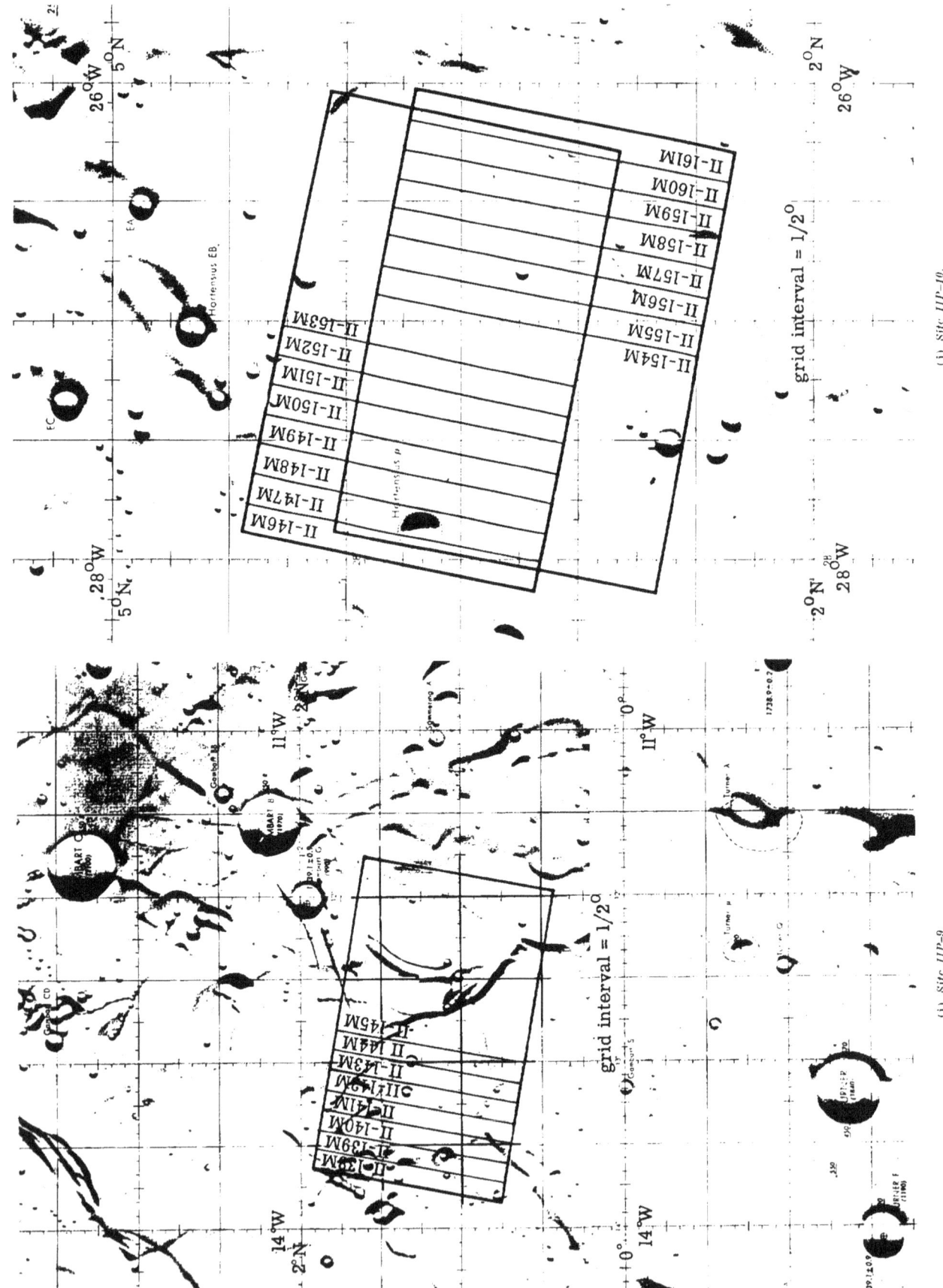

grid interval = 1/2°

(j) Site IIIP-10.

grid interval = 1/2°

(i) Site IIIP-9.

FIGURE 13.—*Photographic Indexes to mission II near-side sites.*—Continued.

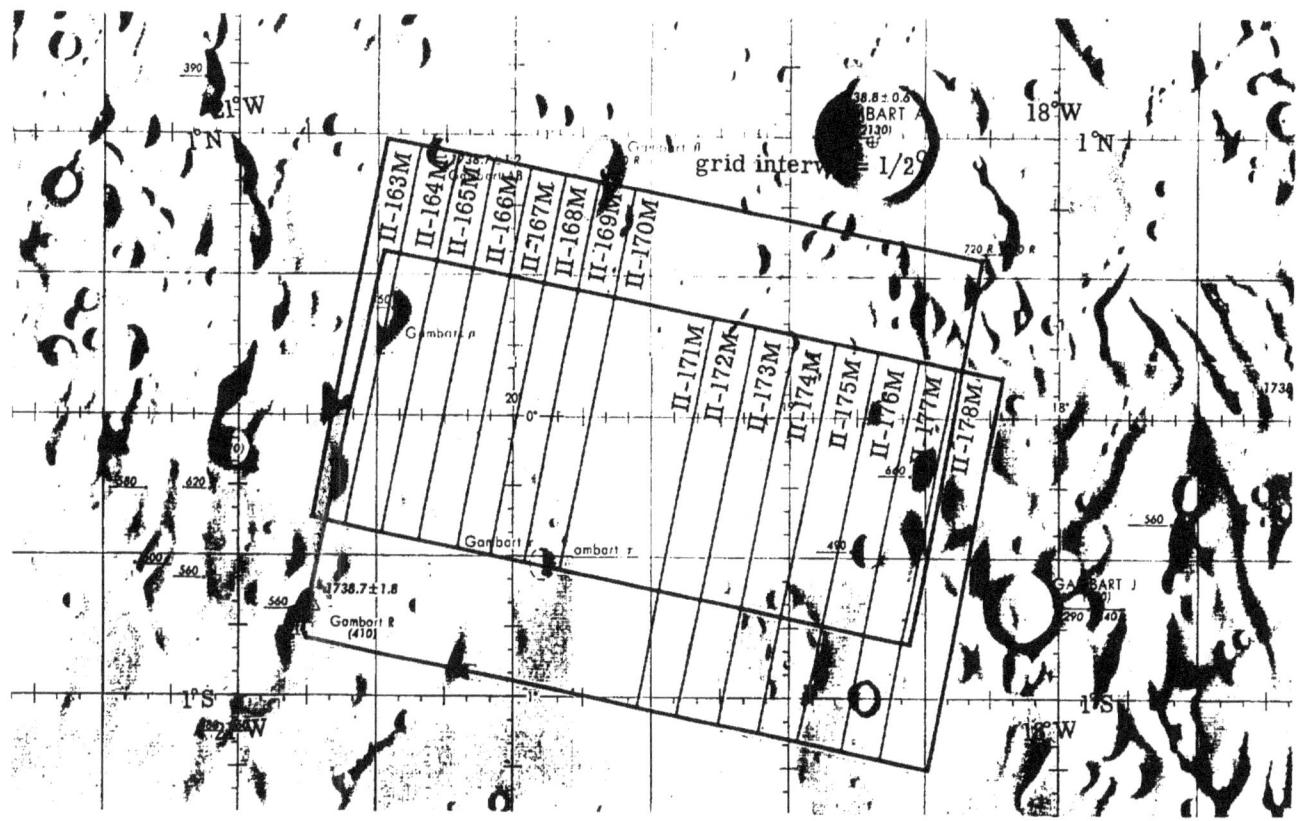

(k) *Site IIP-11.*

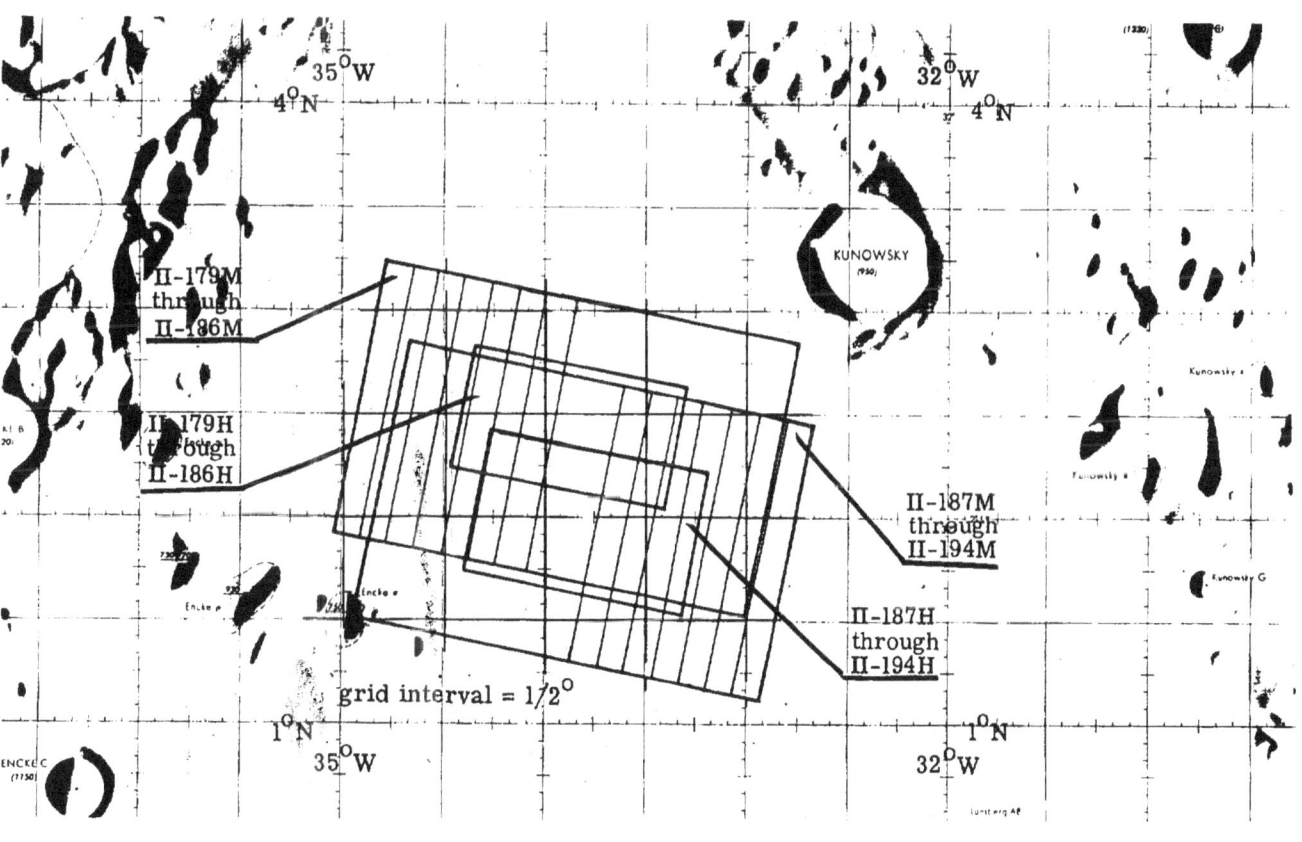

(l) *Site IIP-12.*

FIGURE 13.—*Photographic Indexes to mission II near-side sites.*—Continued.

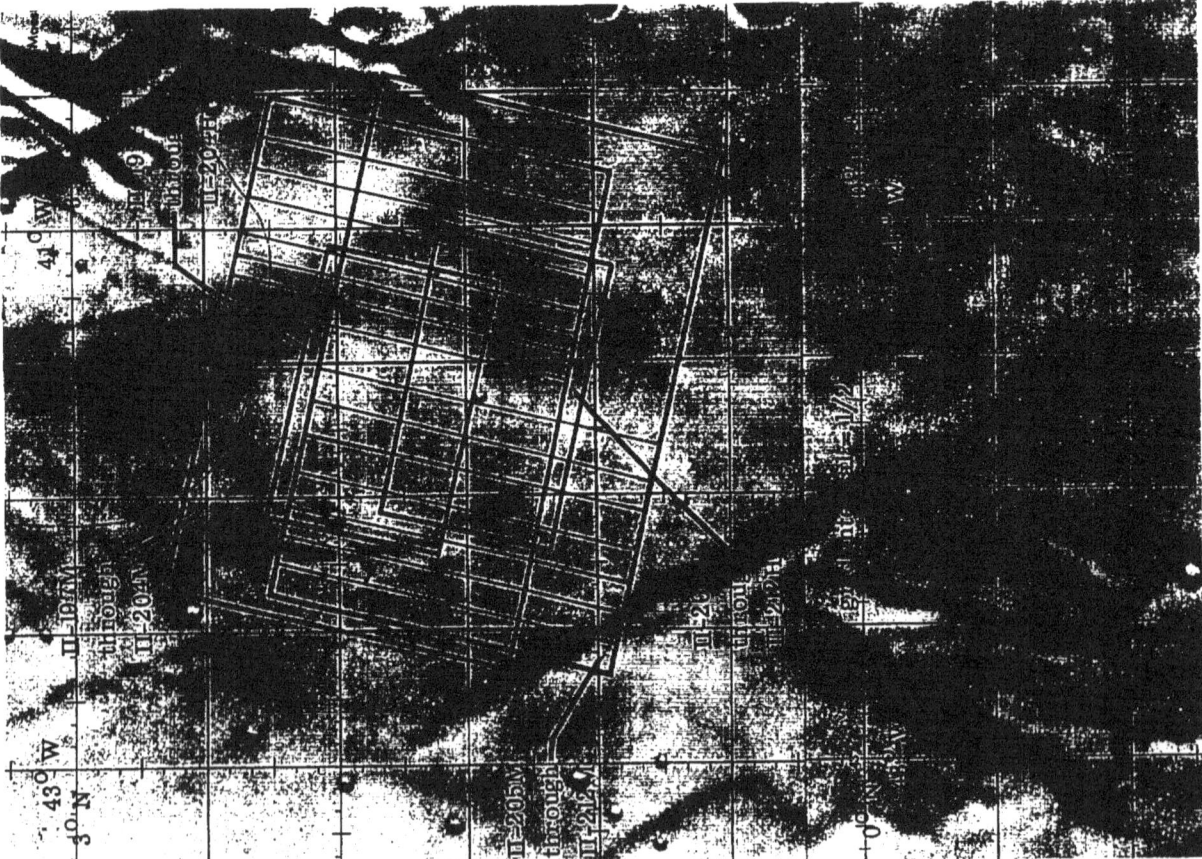

(n) *Site 11S-1.*

(m) *Site 11P-13.*

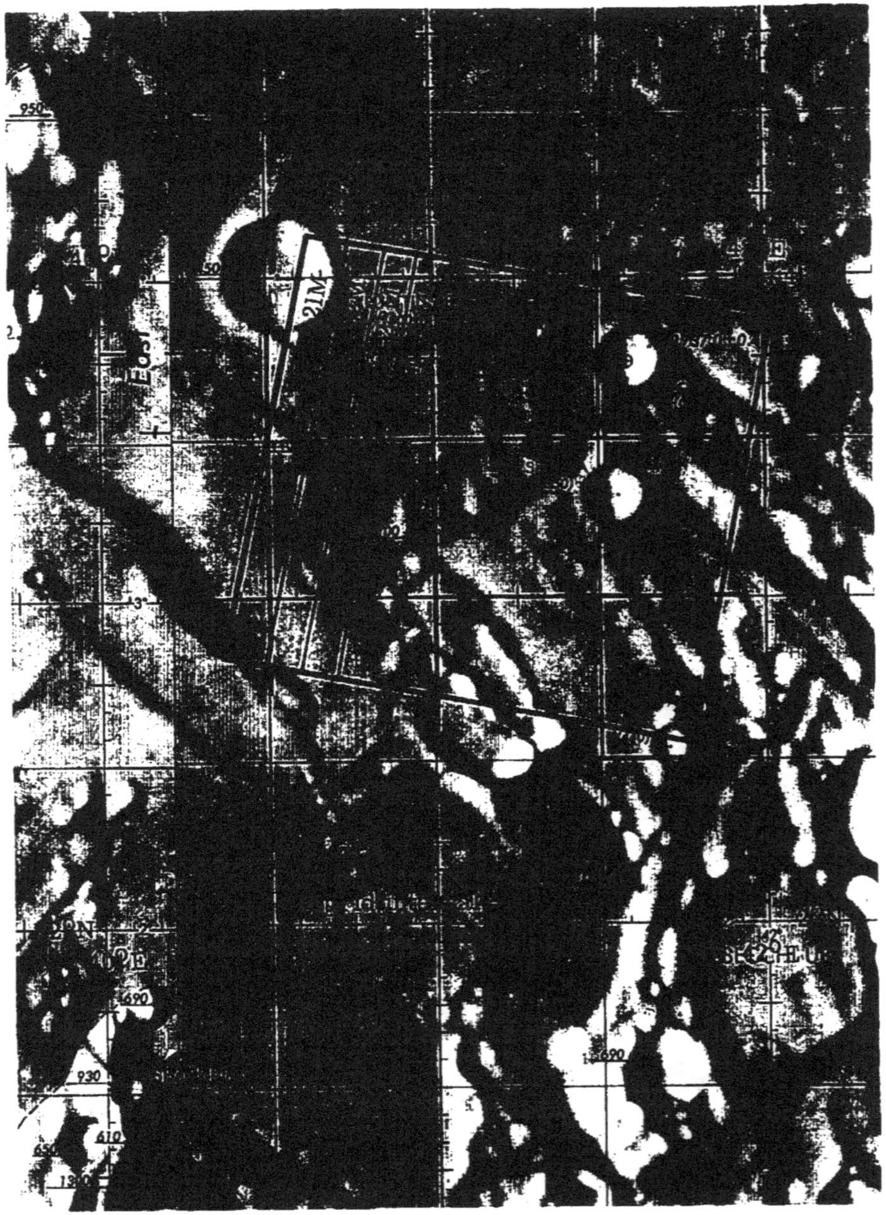

FIGURE 13.—*Photographic Indexes to mission 11 near-side sites.*—Continued.

(b) Site IIS-6.

FIGURE 13.—Photographic Indexes to mission II near-side sites.—Continued.

(a) Site IIS-2.

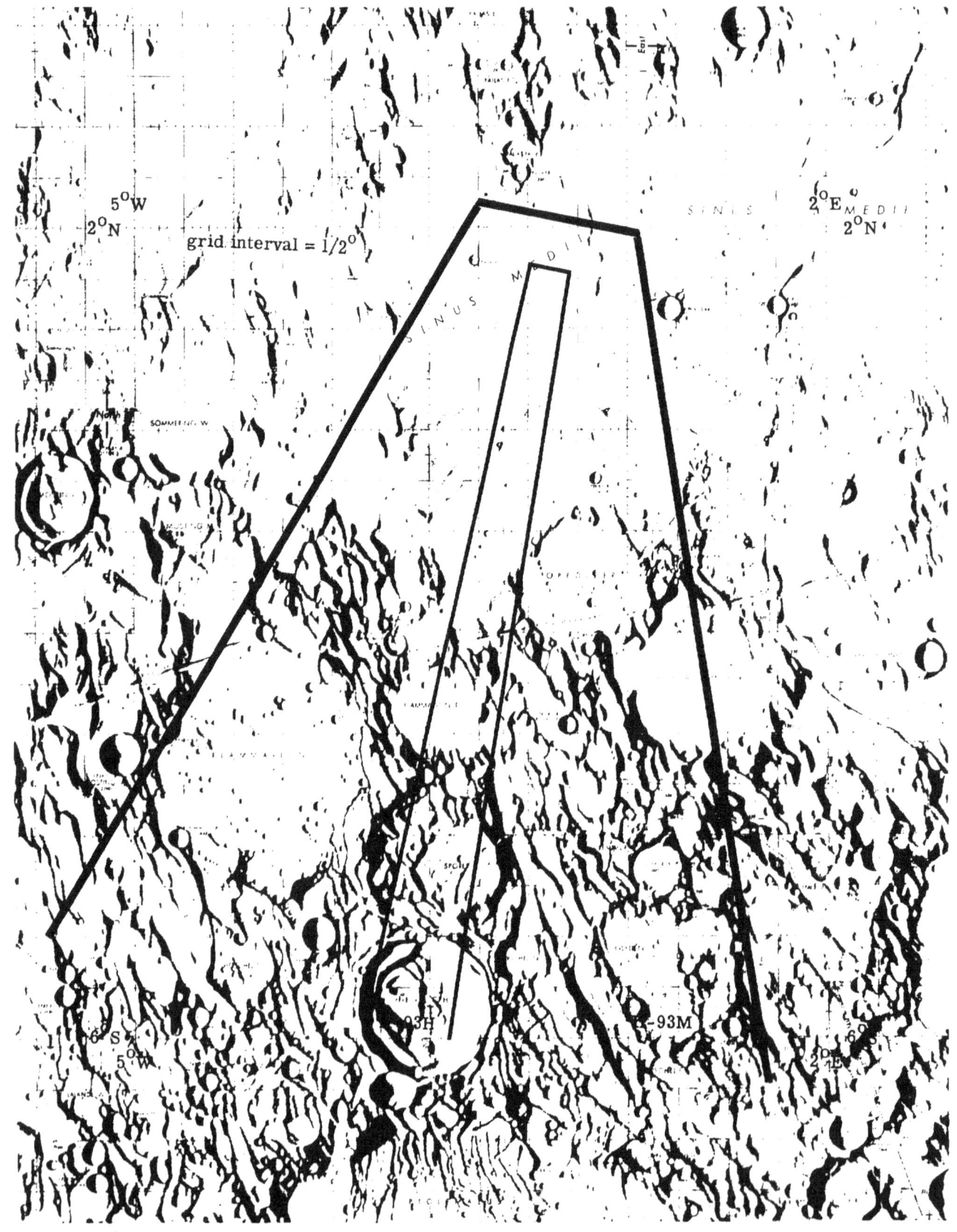

(q) *Site 118-7.*

FIGURE 13.—*Photographic Indexes to mission II near-side sites.*—Continued.

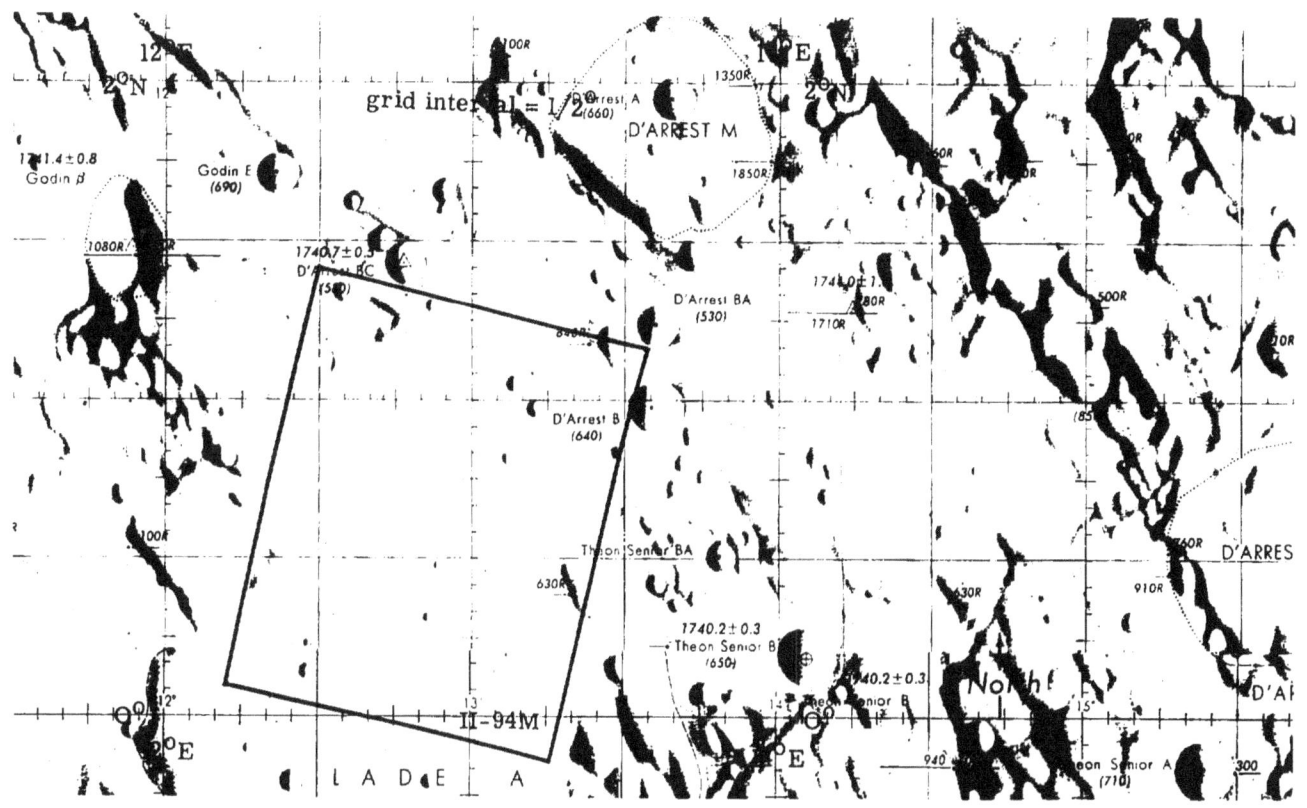

(r) Site IIS-8.

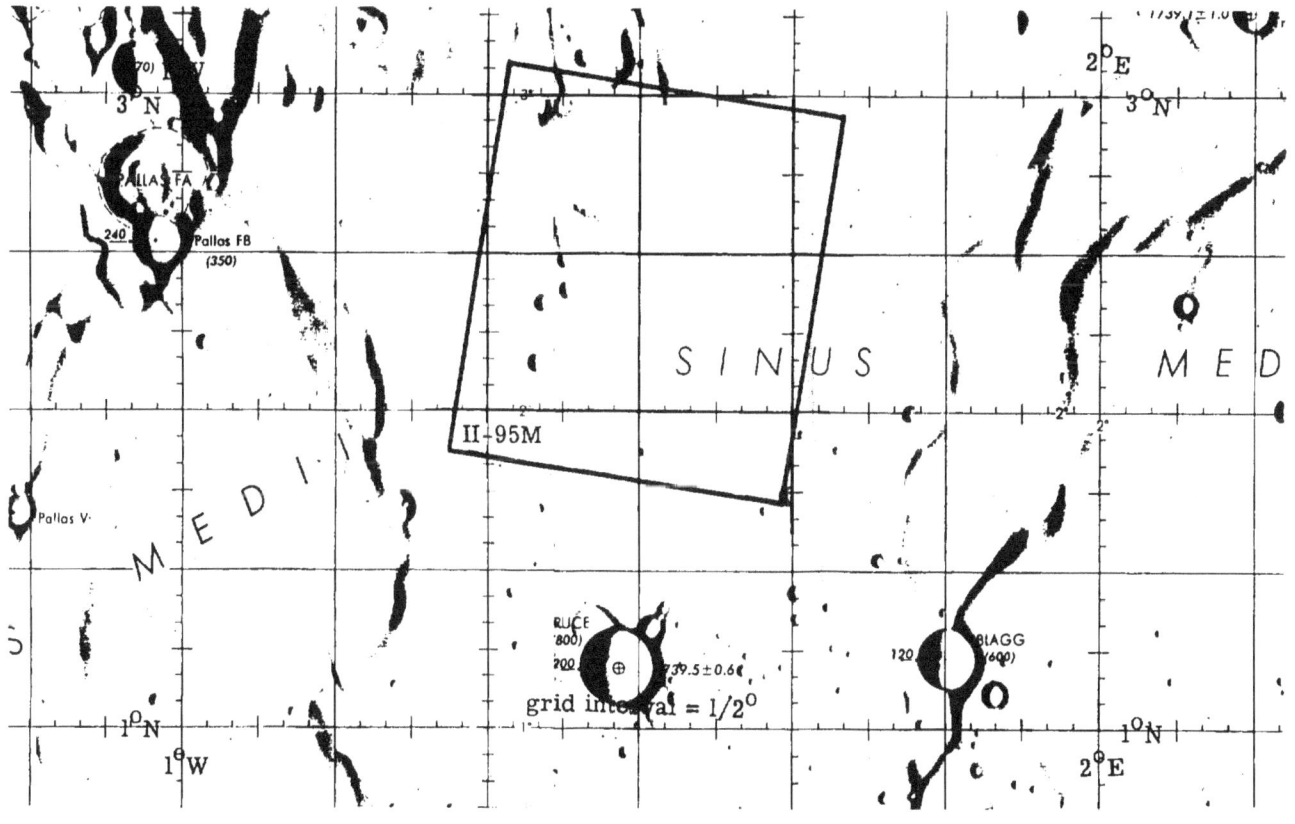

(s) Site IIS-9.

FIGURE 13.—*Photographic Indexes to mission II near-side sites.*—Continued.

(u) *Site 11S-11.*

(t) *Site 11S-10.2.*

FIGURE 13.—*Photographic Indexes to mission 11 near-side sites.*—Continued.

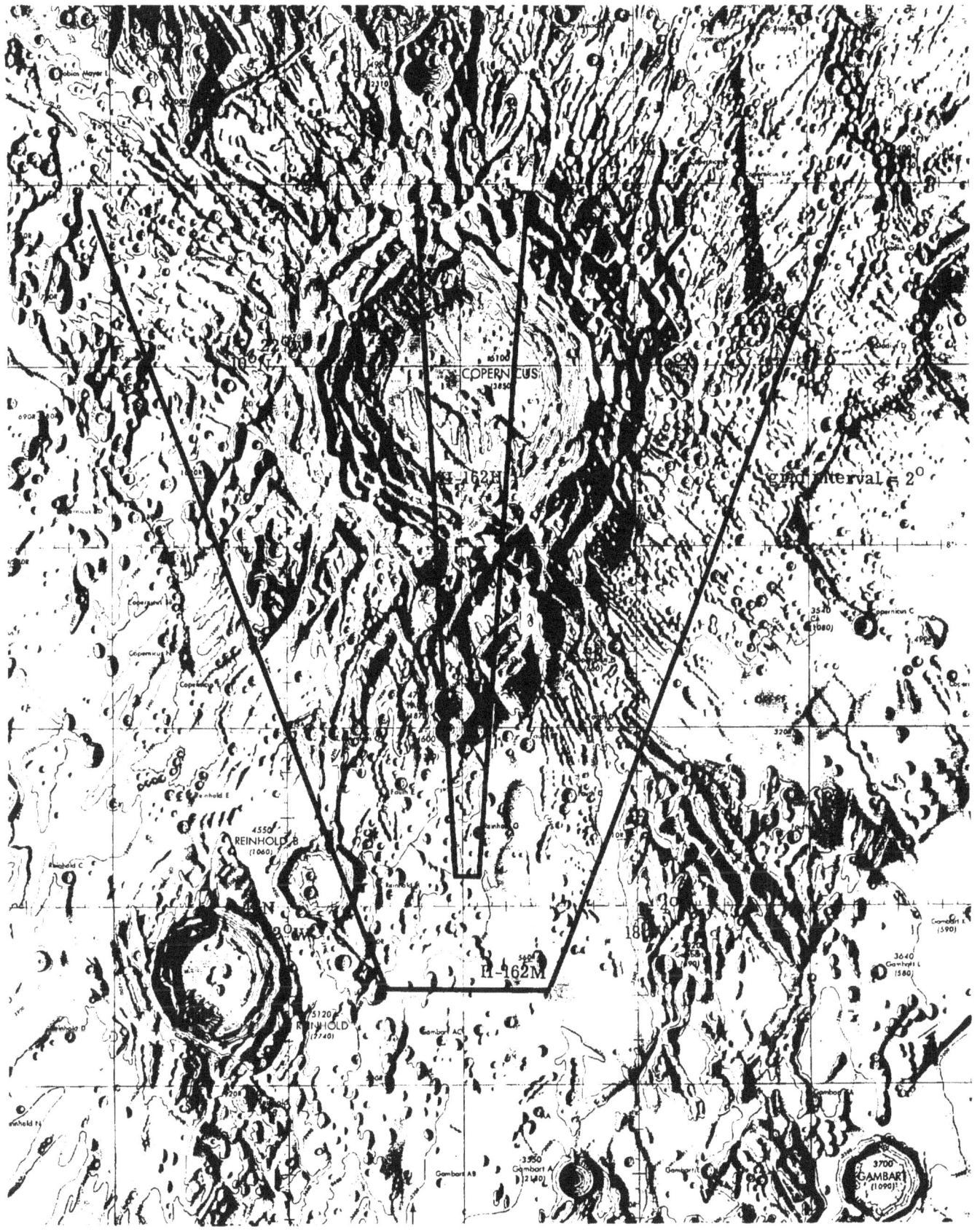

(v) *Site IIS–12.*

FIGURE 13.—*Photographic Indexes to mission II near-side sites.*—Continued.

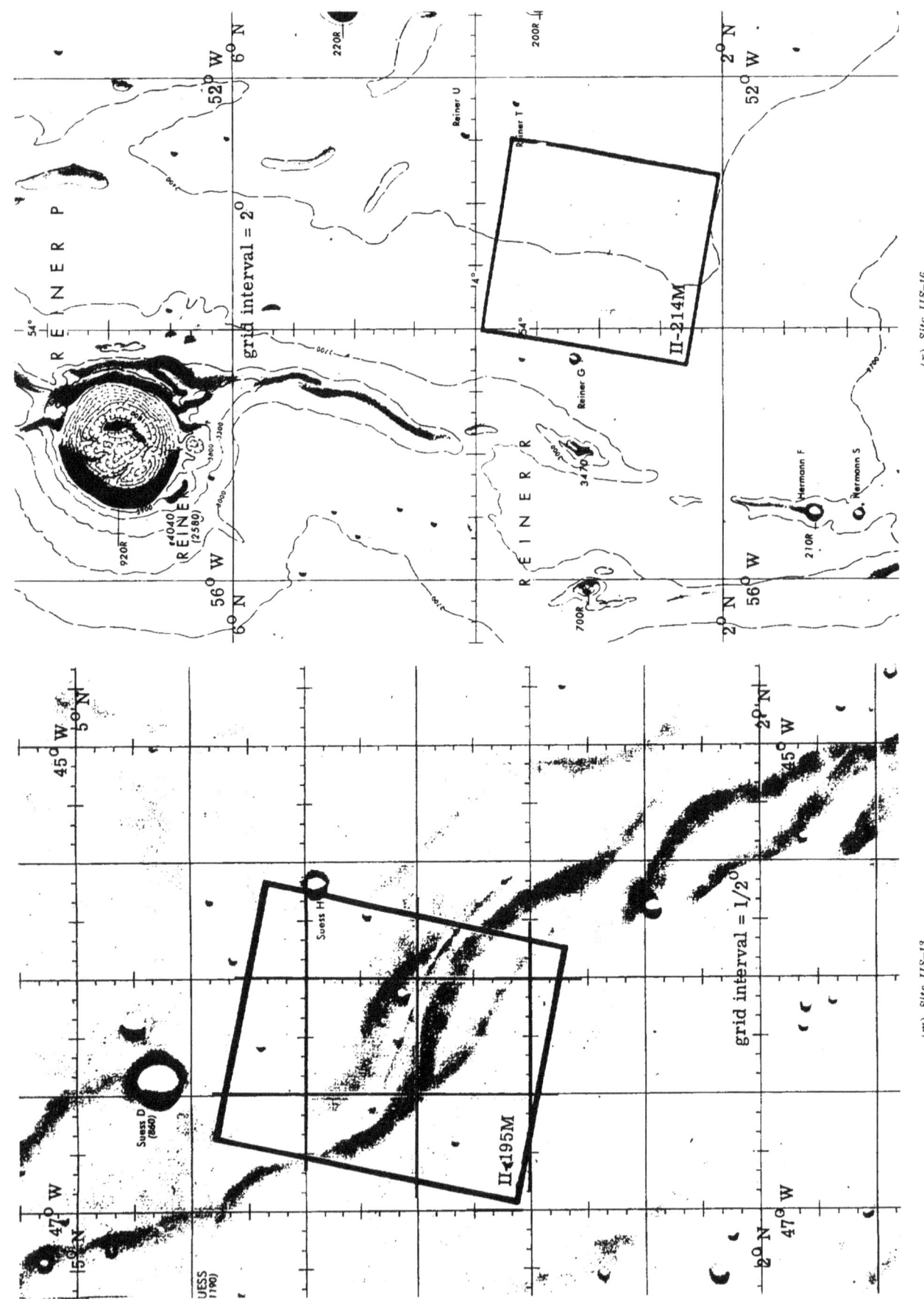

grid interval = 2°

REINER P

54°

REINER

REINER (2580)

₄₀₄₀ (2580)

920R

56° W

6° N

6° N

52° W

220R

Reiner U

Reiner T

200R

2° N

54°

REINER R

Reiner G

3470

Hermann F

Hermann S

210R

700R

56° W

2° N

II-214M

52° W

2° N

(x) *Site IIS-16.*

grid interval = 1/2°

45° W

5° N

47° W

5° N

Suess D (860)

Suess H

UESS (1190)

II-195M

47° W

2° N

2° N

45° W

2° N

(w) *Site IIS-13.*

FIGURE 13.—*Photographic Indexes to mission II near-side sites.*—Continued.

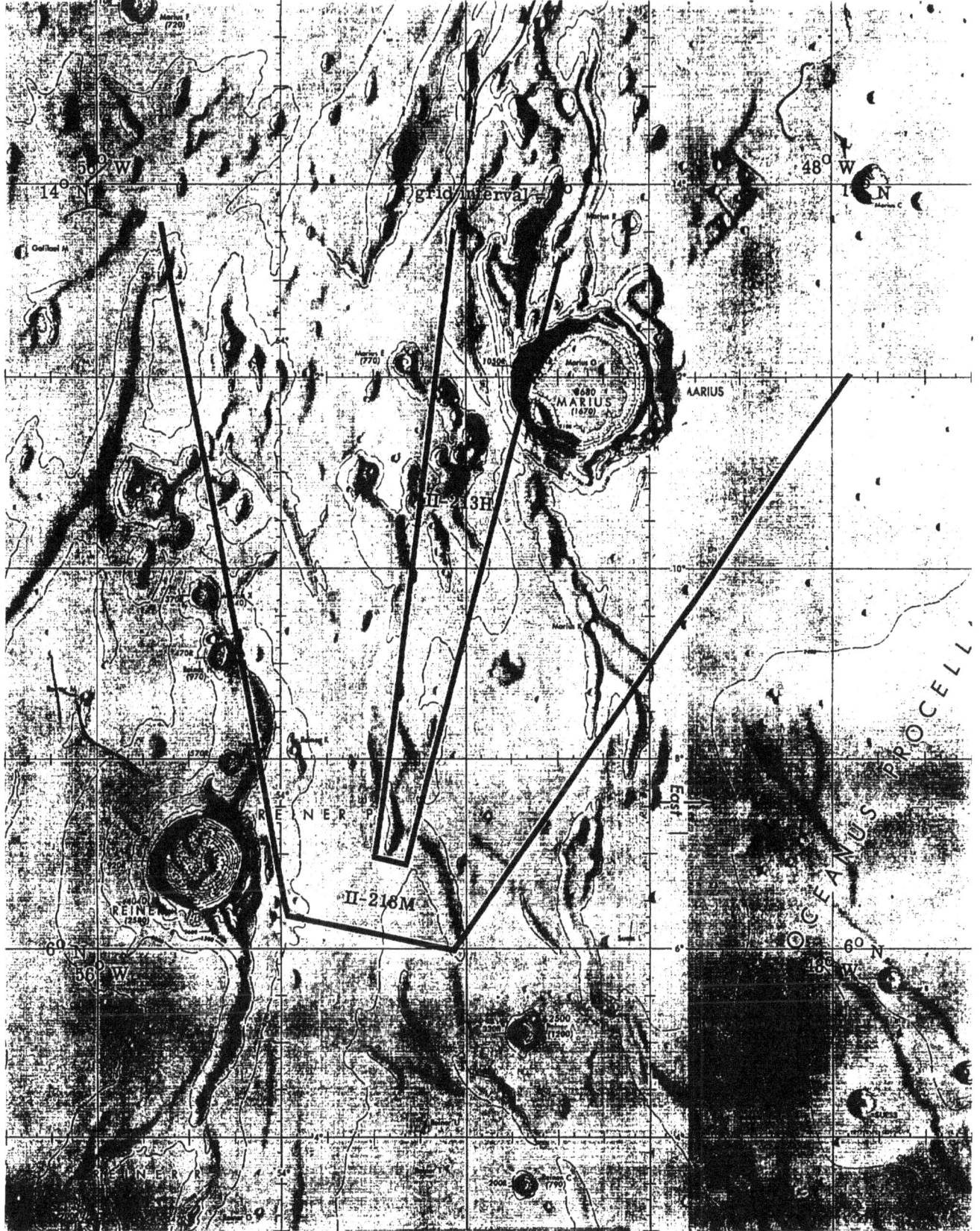

(y) Site IIS-15.

FIGURE 13.—Photographic Indexes to mission II near-side sites.—Continued.

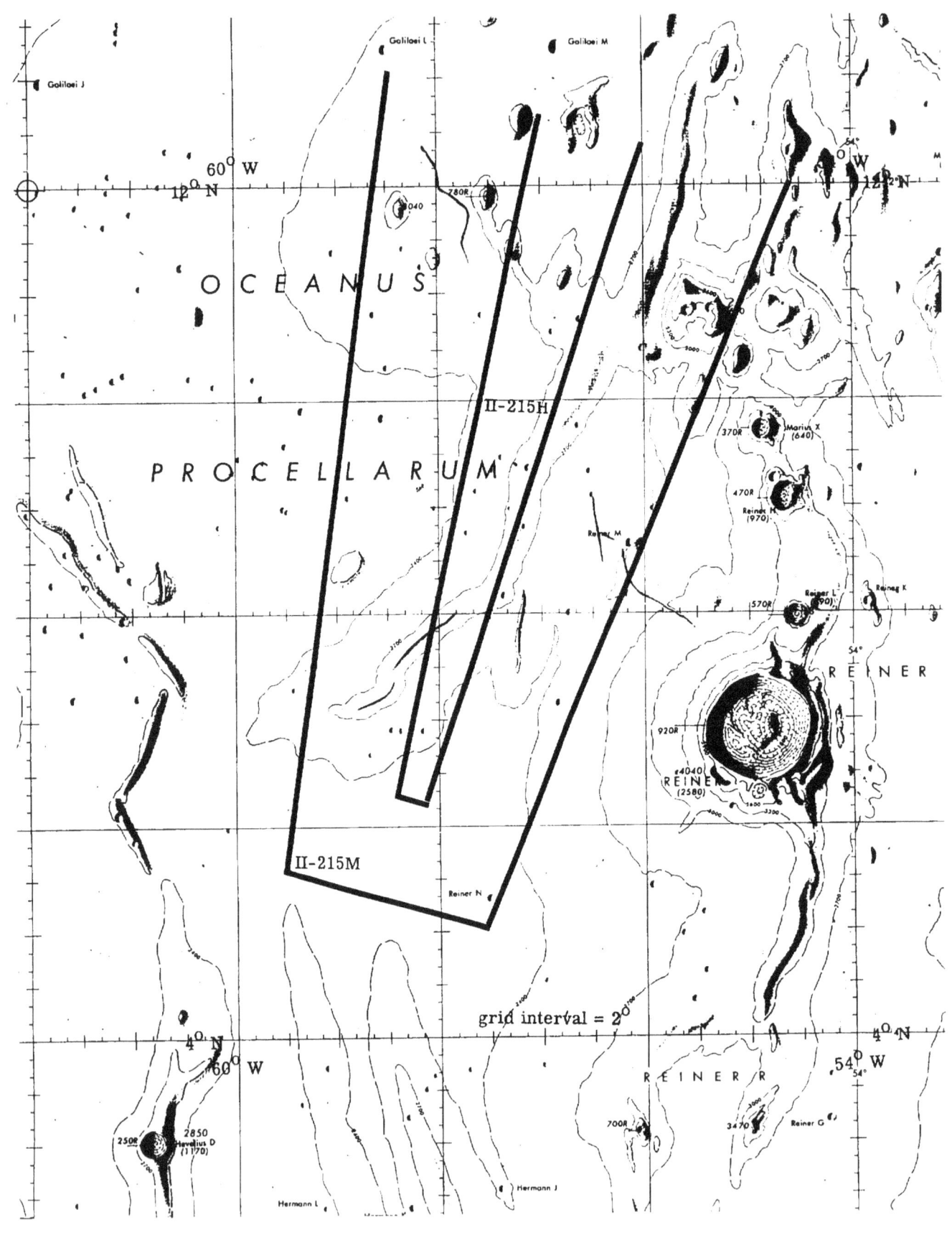

(z) *Site 118-17.*

FIGURE 13.—*Photographic Indexes to mission II near-side sites.*—Concluded.

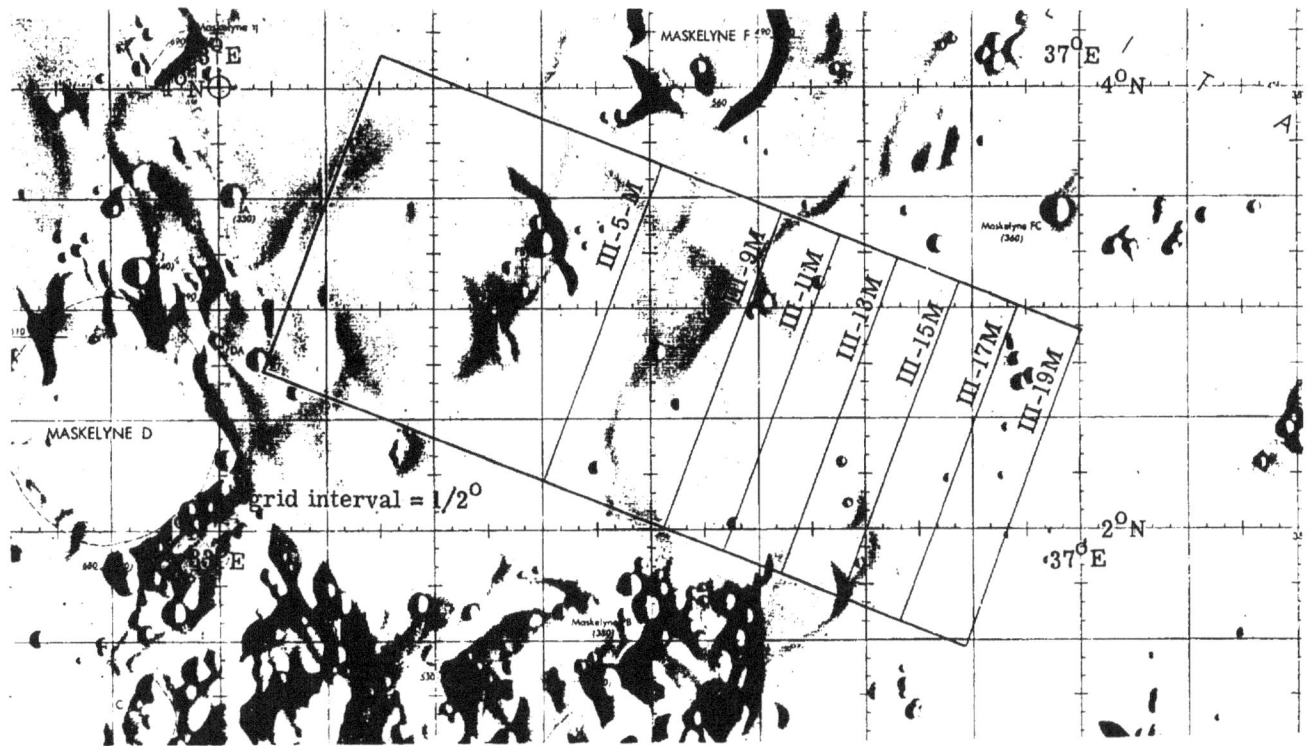

(a) Site IIIP-1.

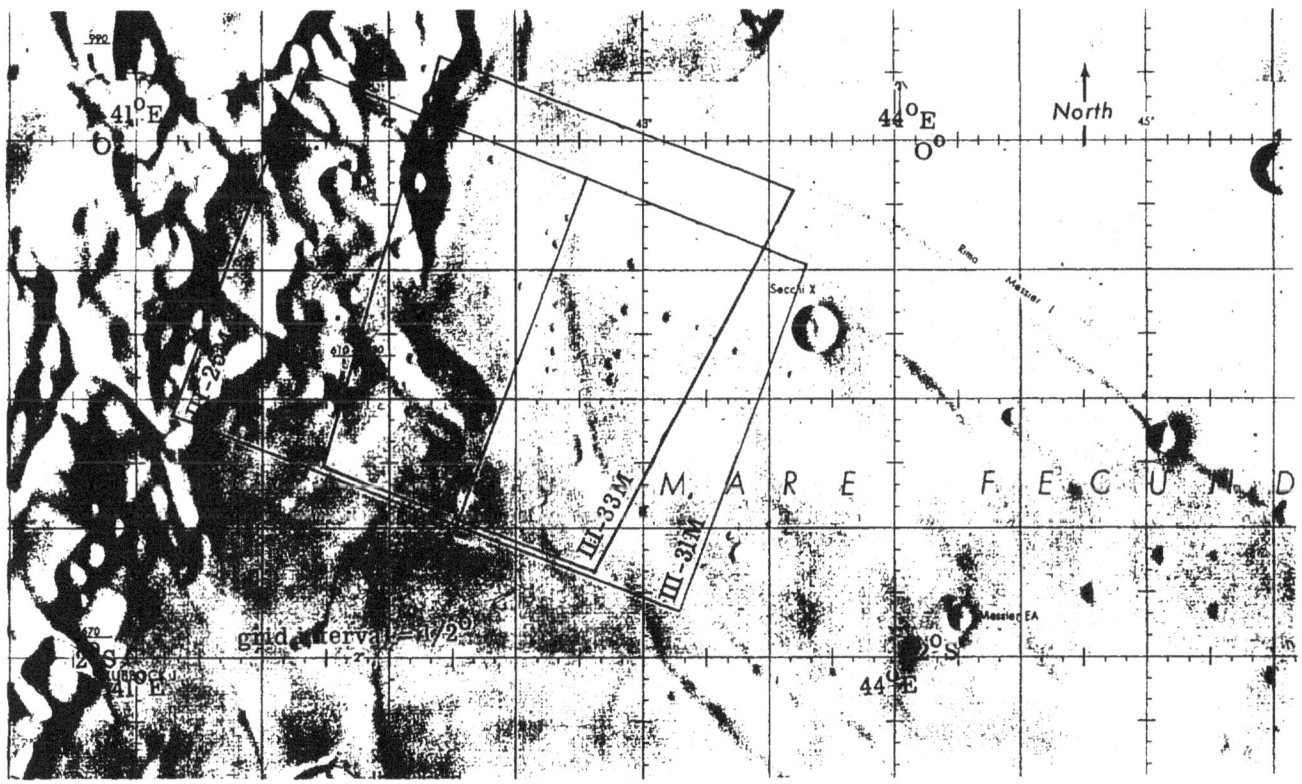

(b) Site IIIP-2.

FIGURE 14.—Photographic Indexes to mission III near-side sites.

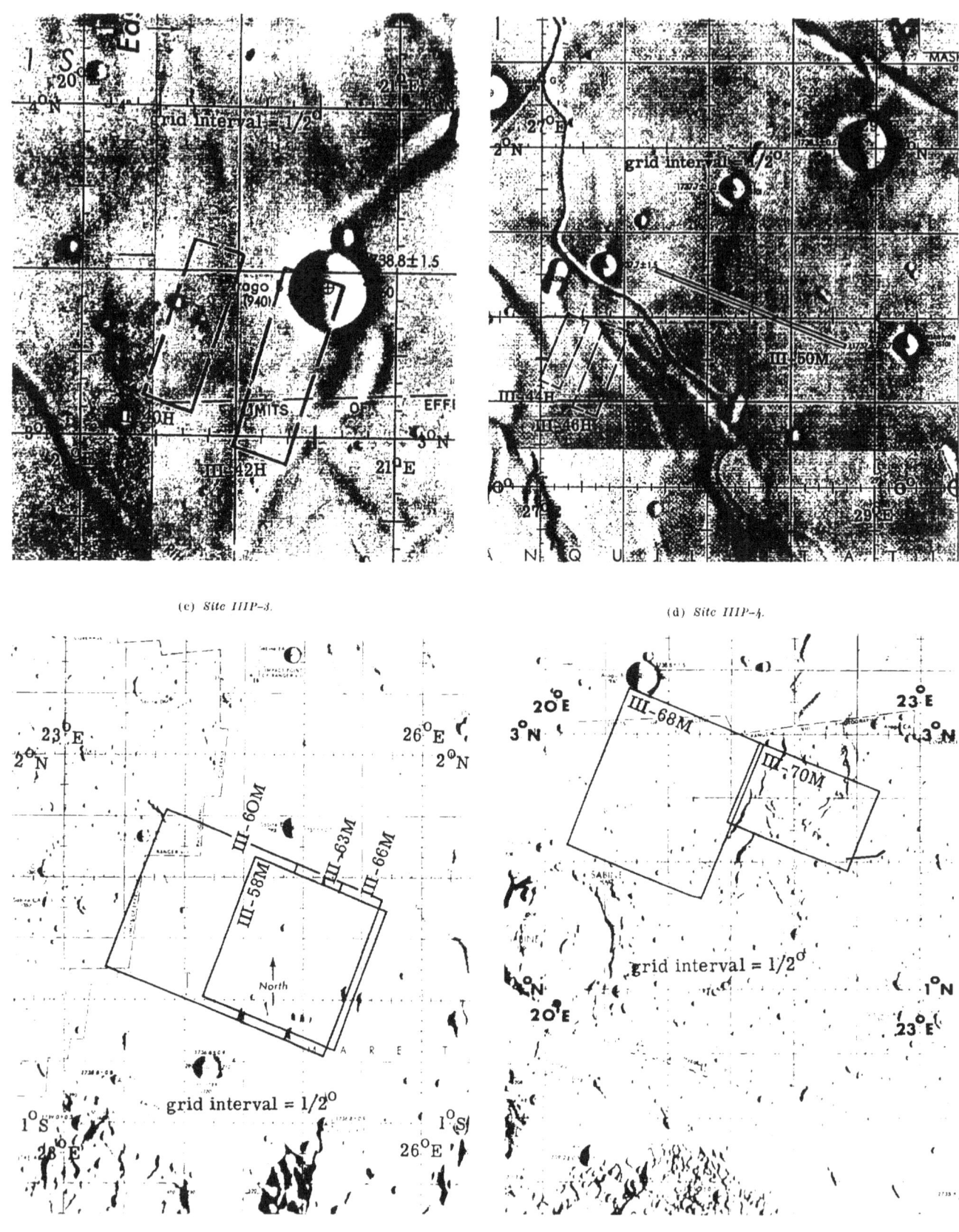

(c) *Site IIIP-3.*

(d) *Site IIIP-4.*

(e) *Site IIIP-5.*

(f) *Site IIIP-6.*

FIGURE 14.—*Photographic Indexes to mission III near-side sites.*—Continued.

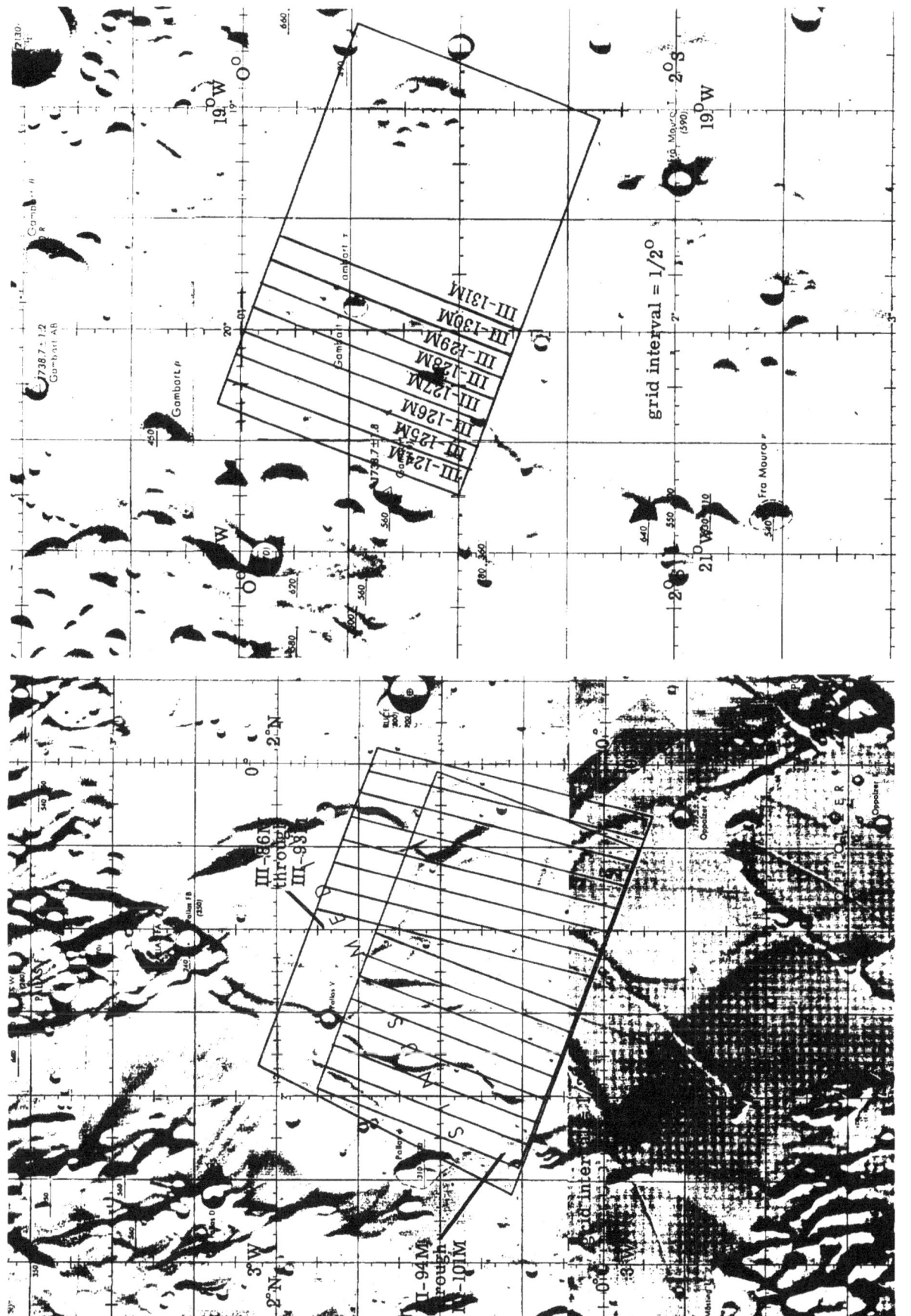

grid interval = 1/2°

III-124M
III-125M
III-126M
III-127M
III-128M
III-129M
III-130M
III-131M

(h) Site IIIP–8.

III-86M through III-88M

III-94M through III-101M

grid interval

(g) Site IIIP–7.

Figure 14.—*Photographic Indexes to mission III near-side sites.*—Continued.

(j) Site IIIP-10.

(1) Site IIIP-9.

FIGURE 14.—Photographic Indexes to mission III near-side sites.—Continued.

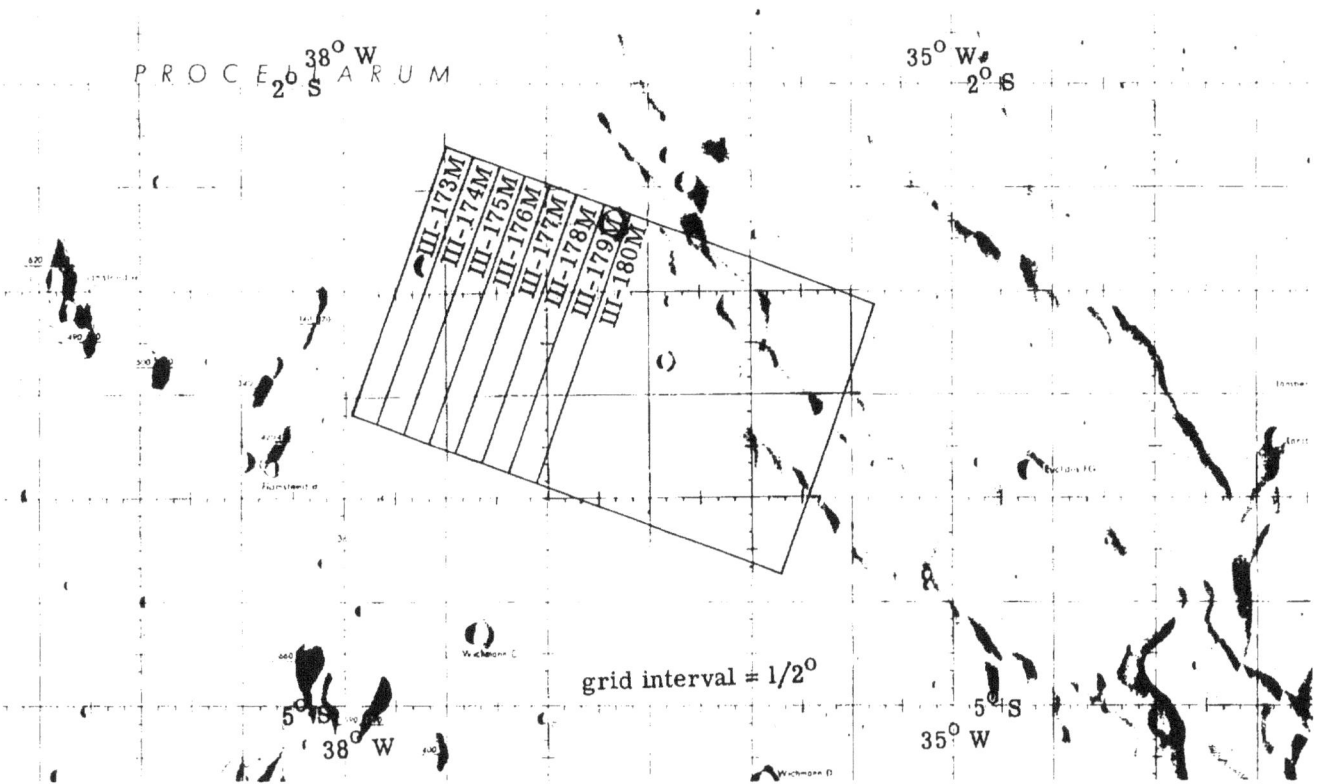

(k) Site IIIP-11.

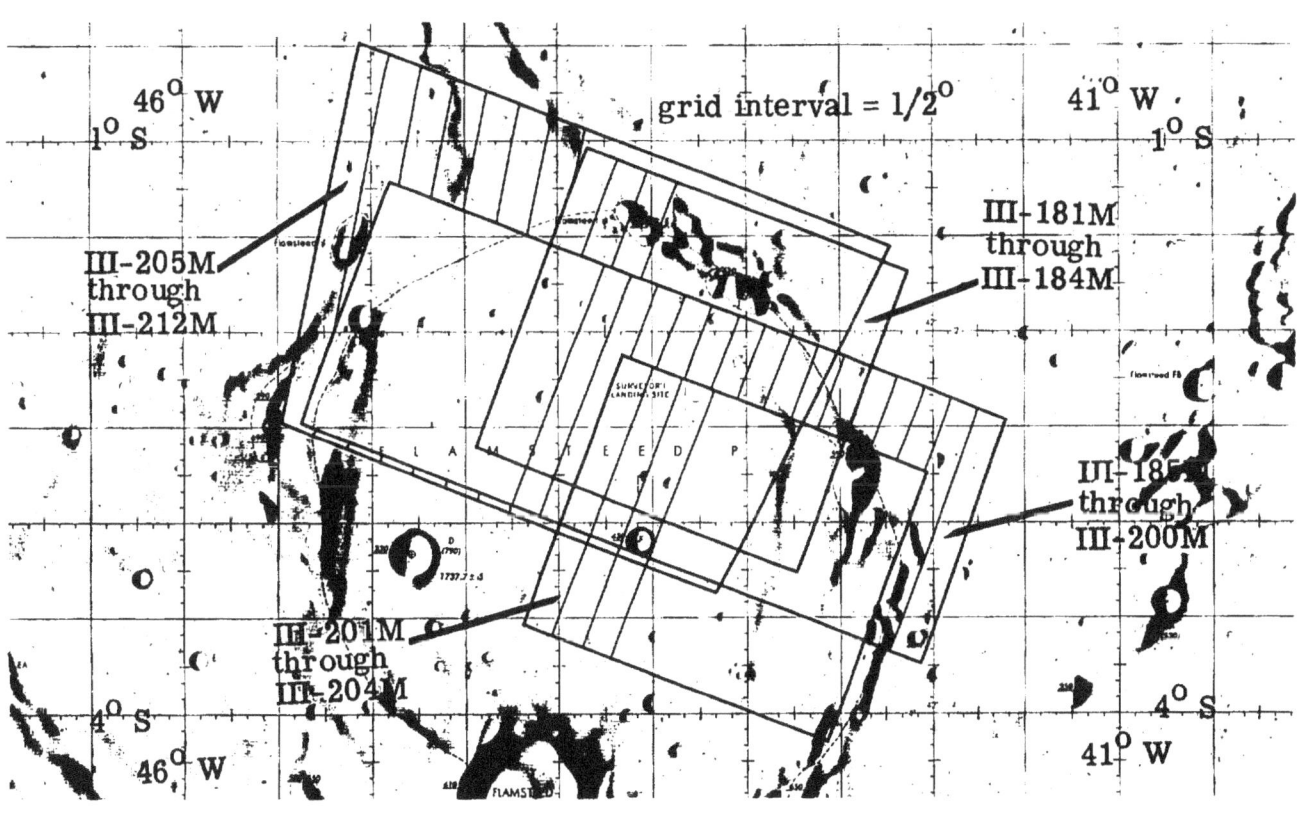

(l) Site IIIP-12.

FIGURE 14.—Photographic Indexes to mission III near-side sites.—Continued.

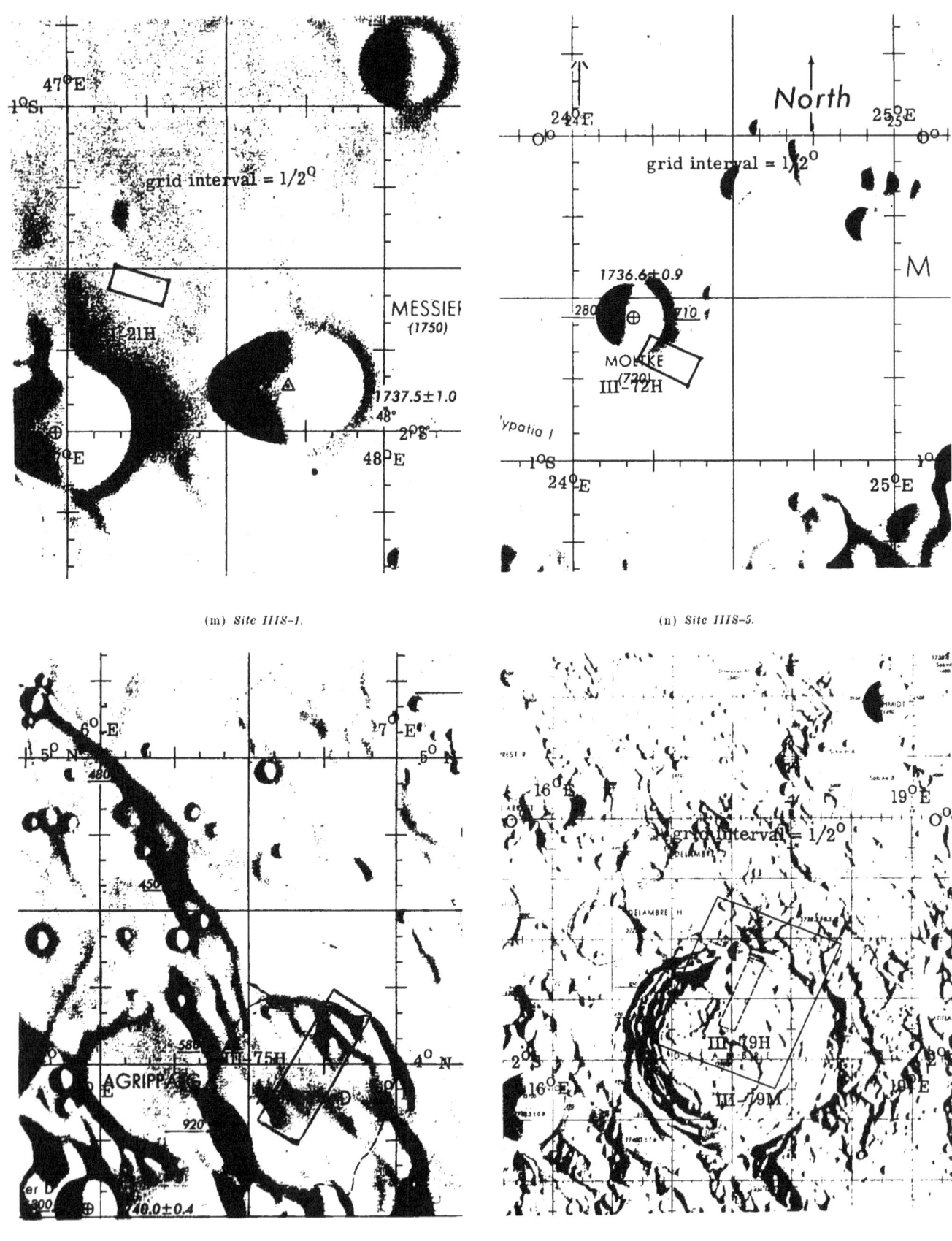

(m) *Site IIIS-1.*

(n) *Site IIIS-5.*

(o) *Site IIIS-7.*

(p) *Site IIIS-9.*

FIGURE 14.—*Photographic Indexes to mission III near-side sites.*—Continued.

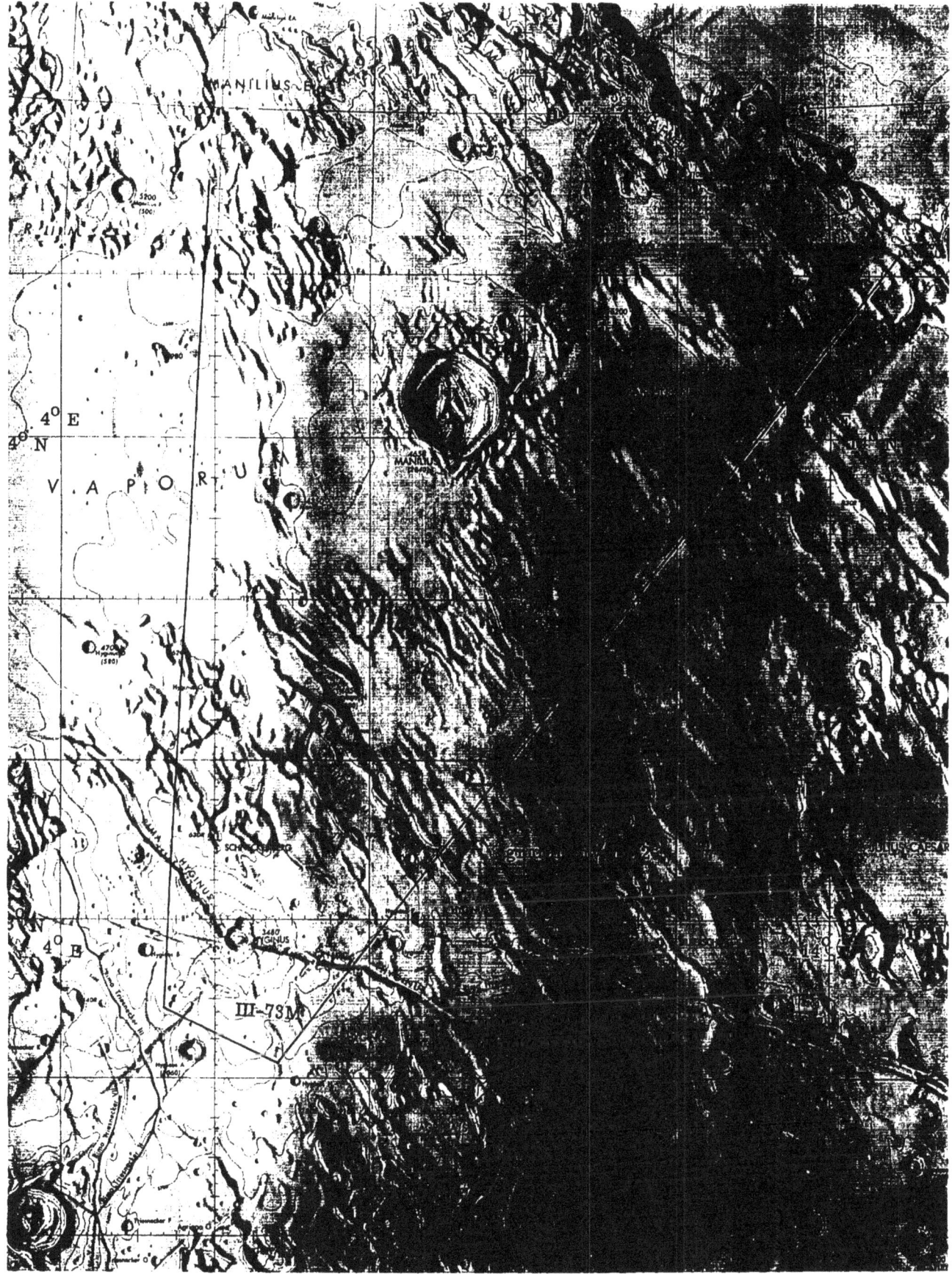

(q) *Site IIIS–6.*

FIGURE 14.—*Photographic Indexes to mission III near-side sites.*—Continued.

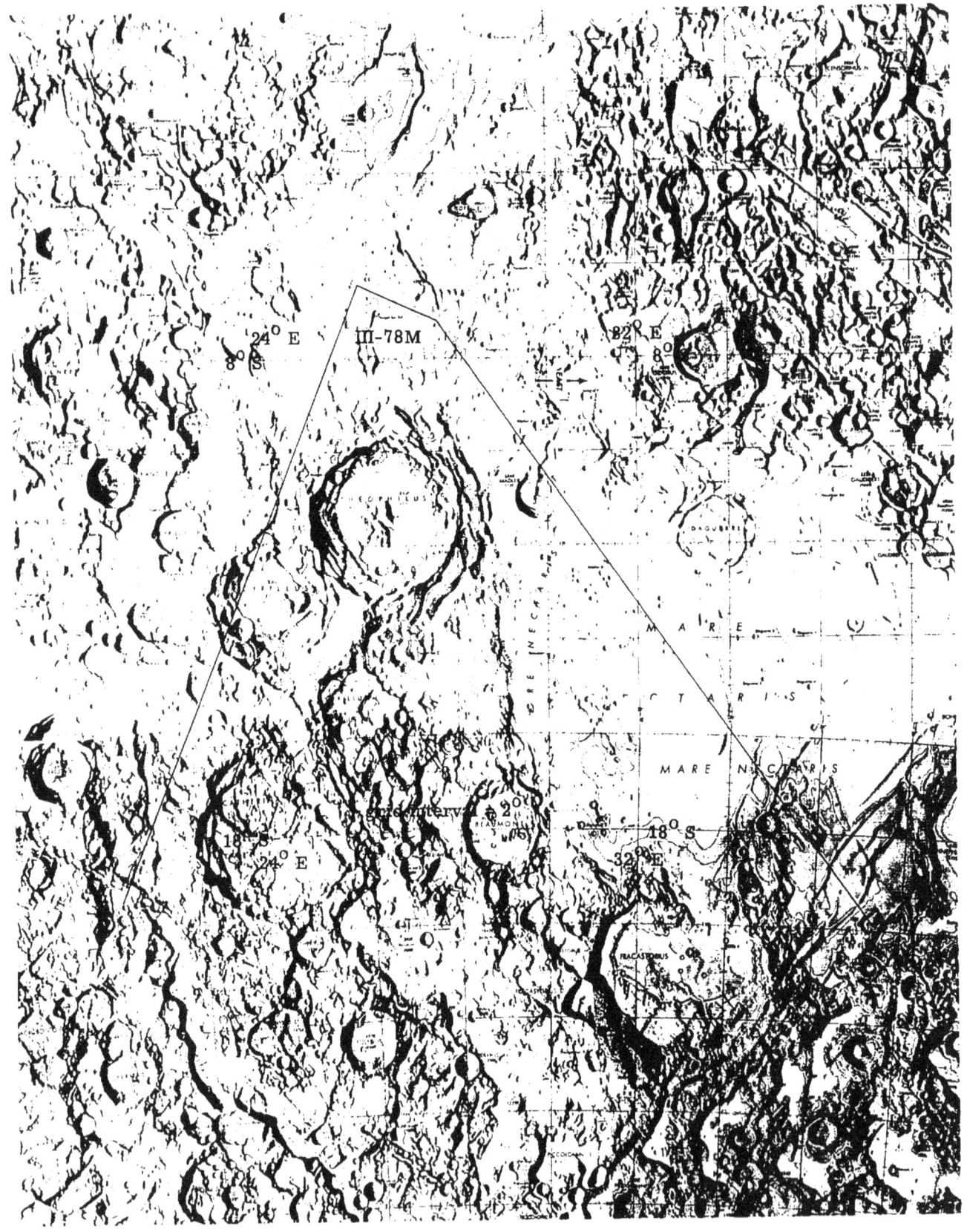

(r) Site IIIS-8.

FIGURE 14.—*Photographic Indexes to mission III near-side sites.*—Continued.

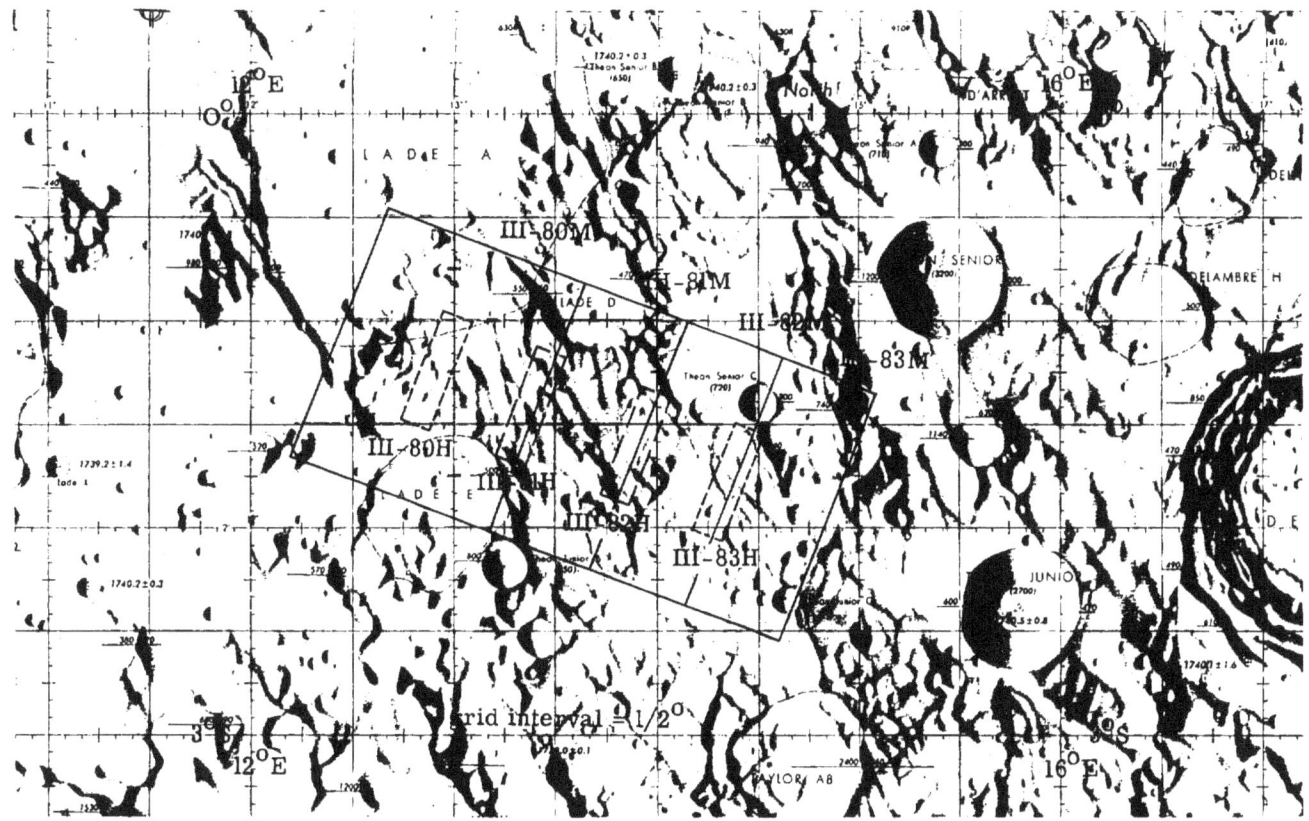

(s) Site IIIS-10.

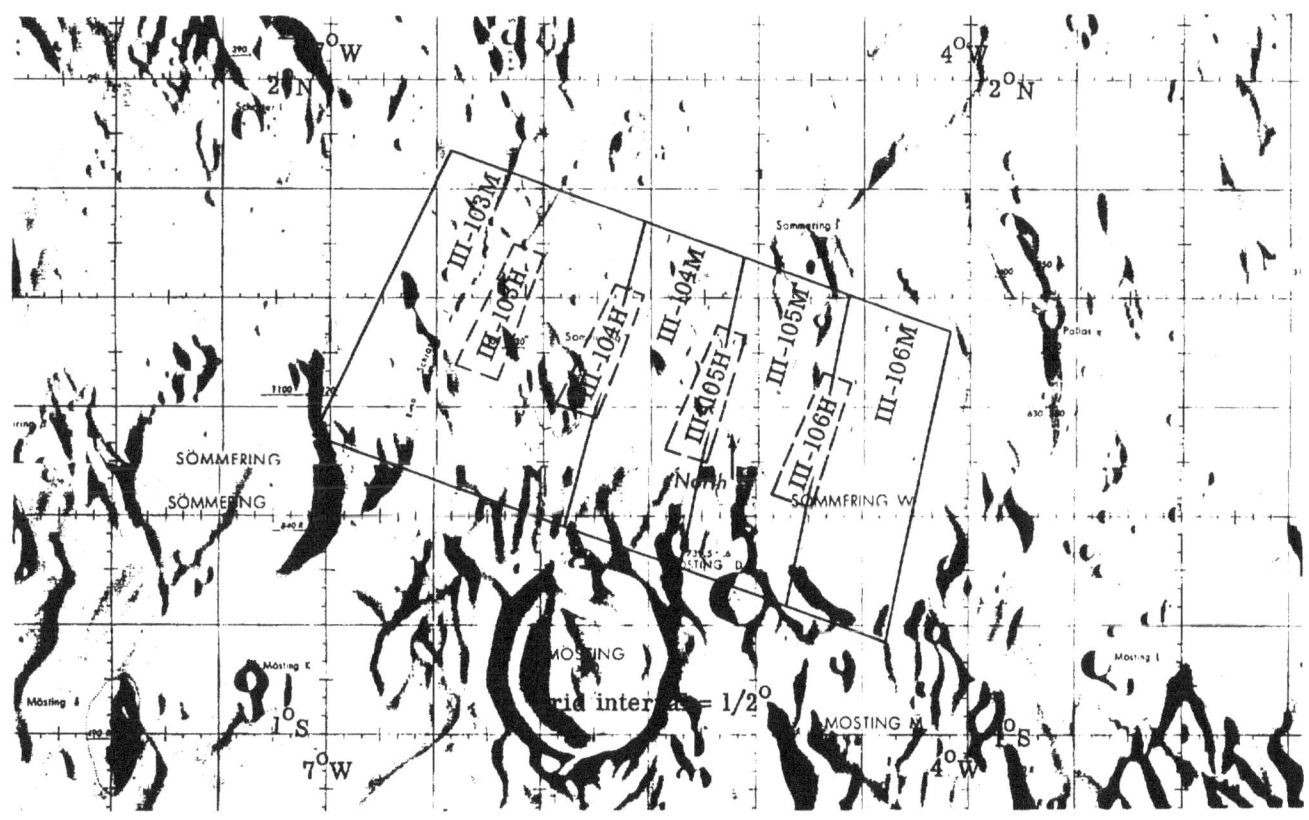

(t) Site IIIS-15.

FIGURE 14.—Photographic Indexes to mission III near-side sites.—Continued.

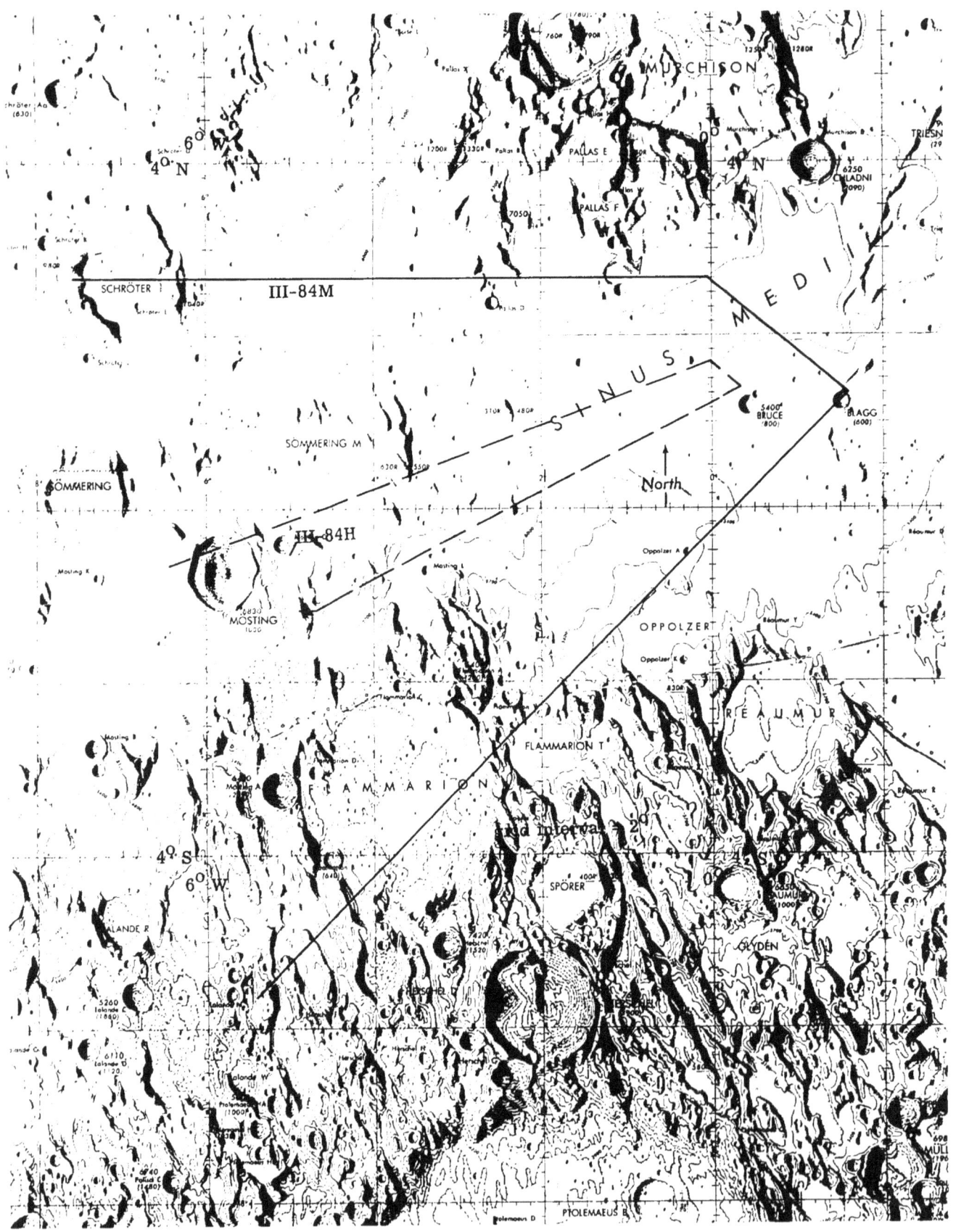

(u) *Site IIIS–11.*

FIGURE 14.—*Photographic Indexes to mission III near-side sites.*—Continued.

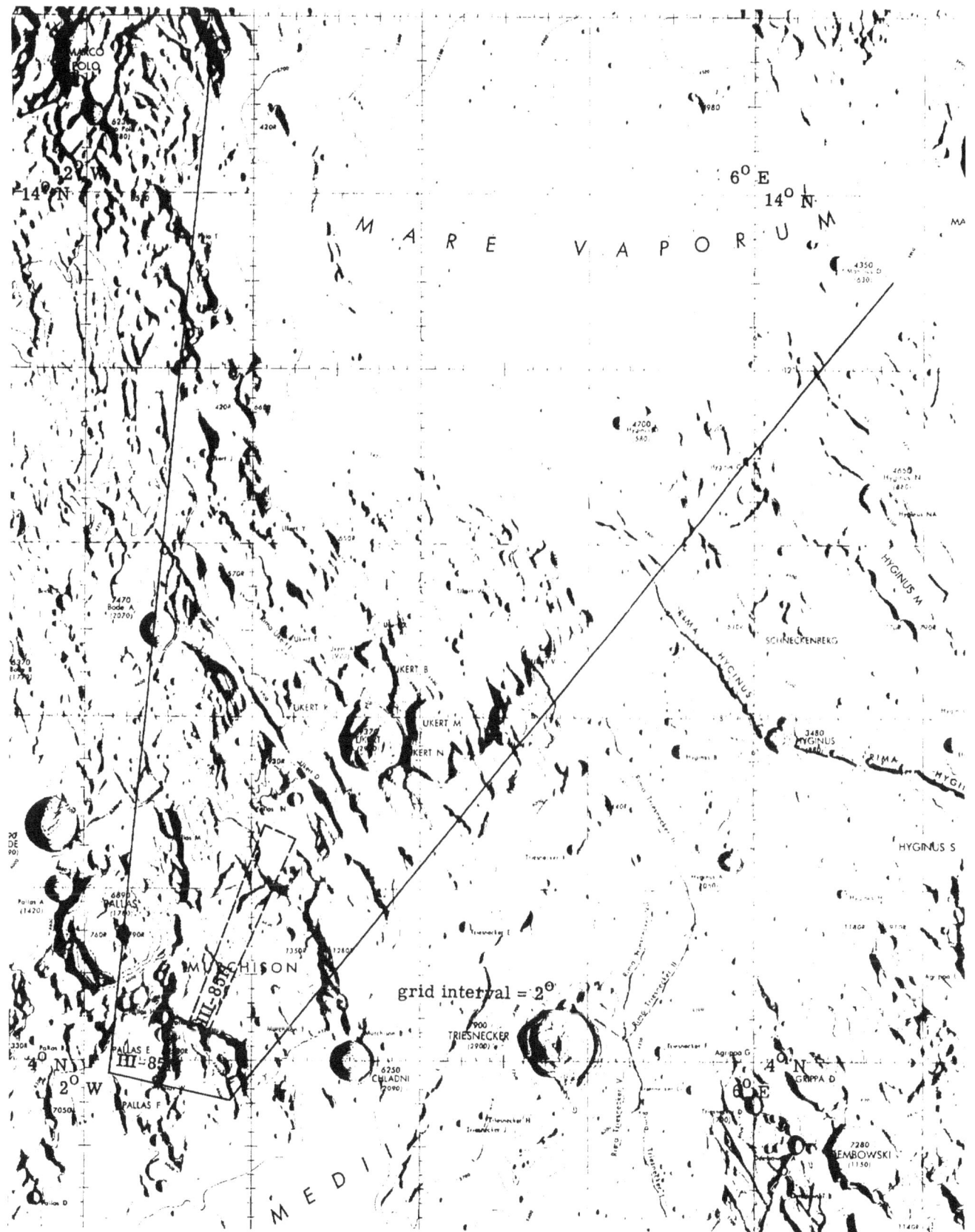

(v) *Site IIIS–13.*

FIGURE 14.—*Photographic Indexes to mission III near-side sites.*—Continued.

(w) *Site IIIS-14.*

FIGURE 14.—*Photographic Indexes to mission III near-side sites.*—Continued.

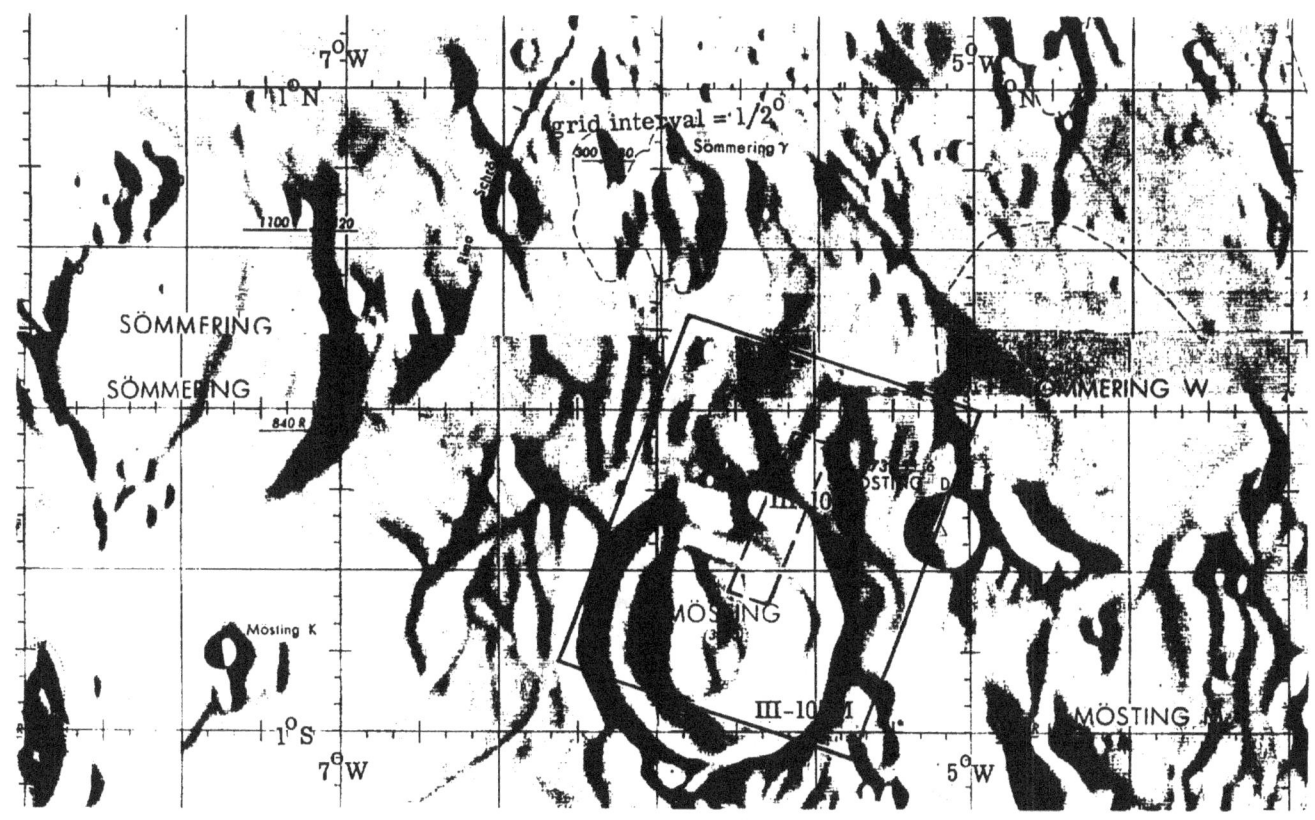

(x) Site IIIS-16.

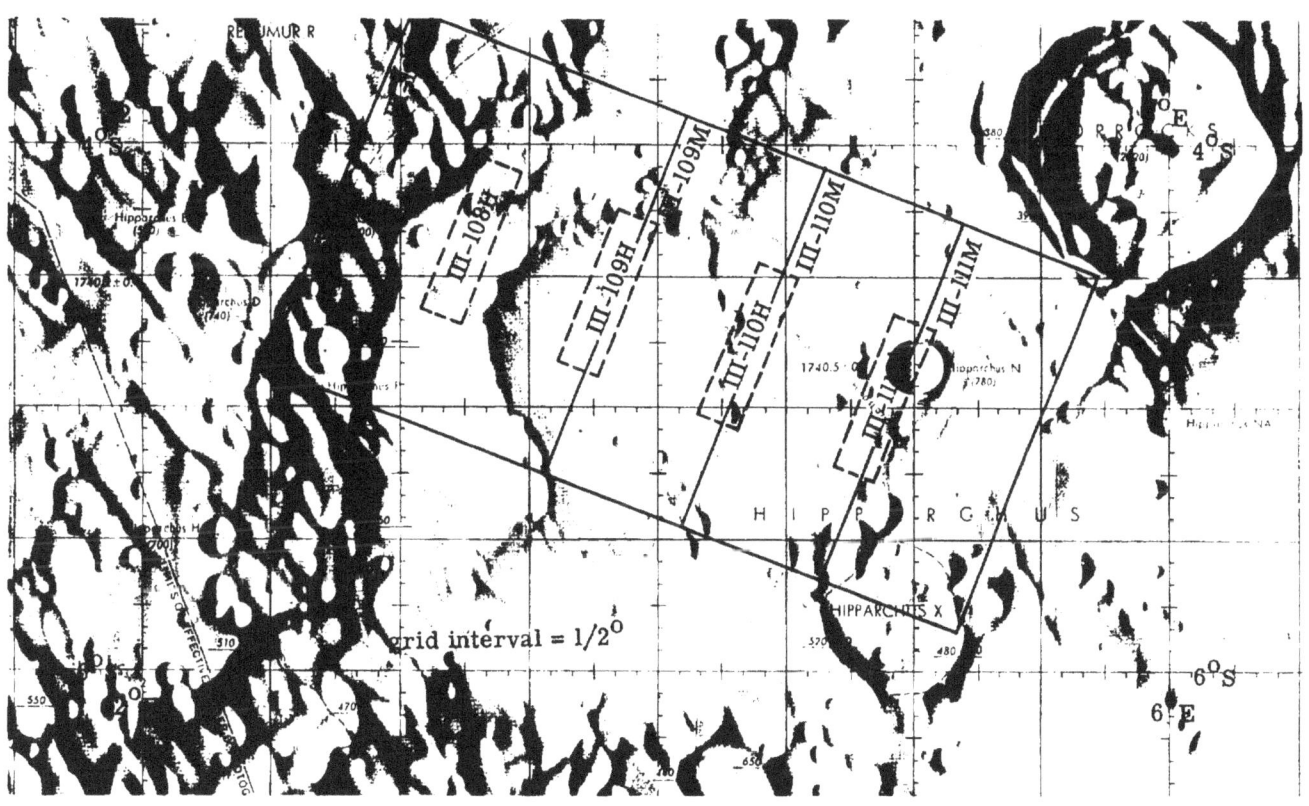

(y) Site IIIS-17.

FIGURE 14.—Photographic Indexes to mission III near-side sites.—Continued.

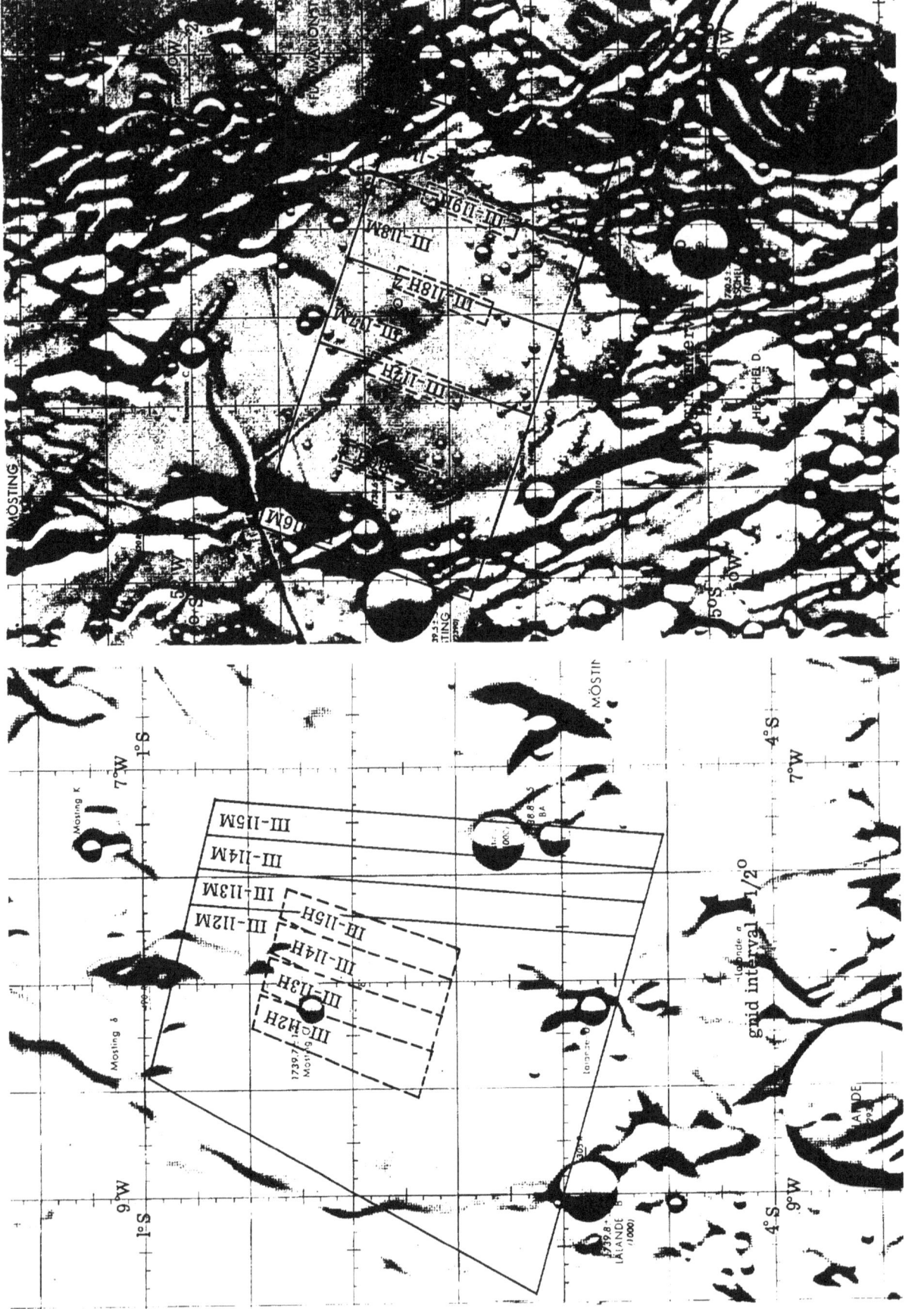

(aa) Site IIIS–19.

(z) Site IIIS–18.

grid interval 1/2°

FIGURE 14.—*Photographic indexes to mission III near-side sites.*—Continued.

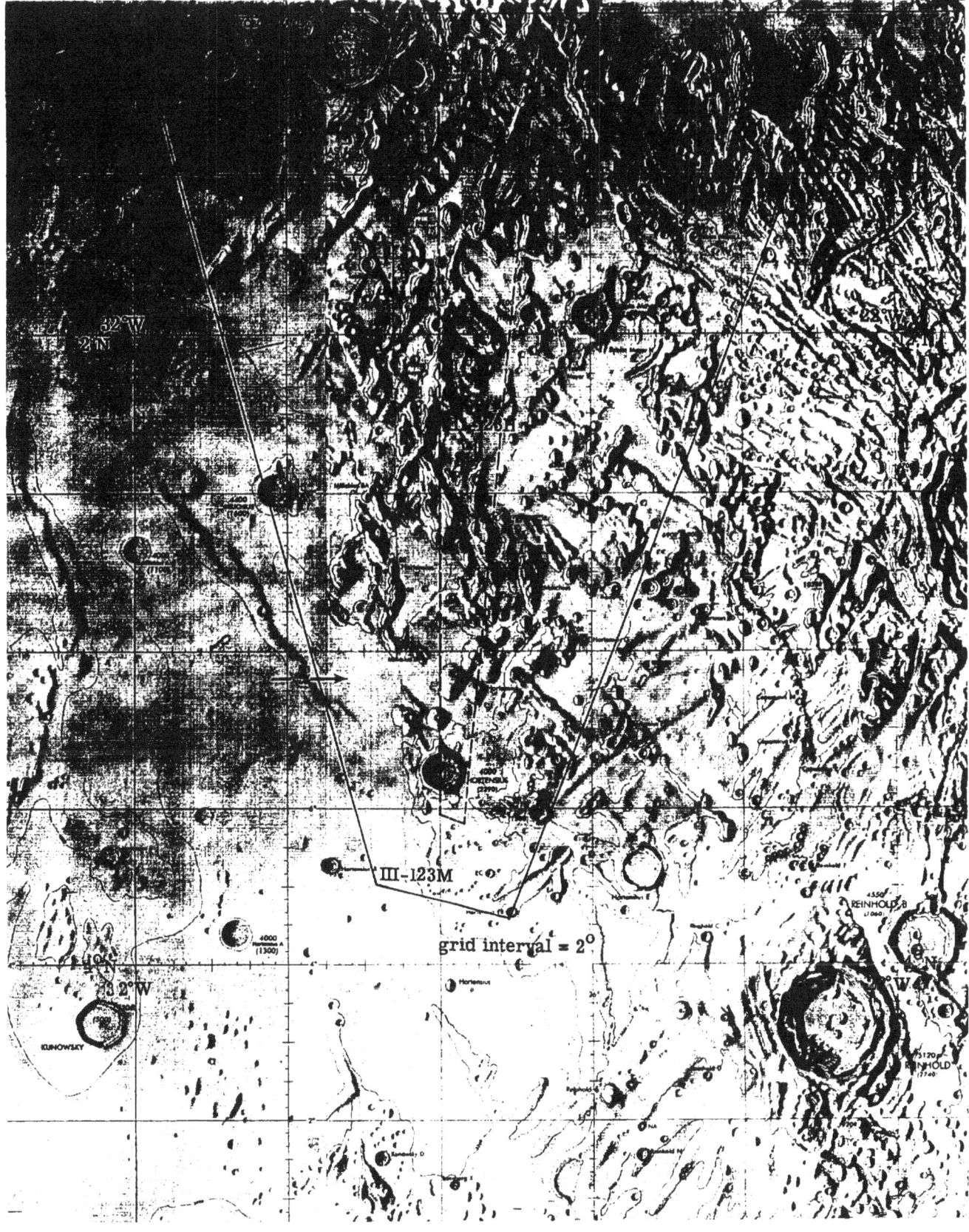

(bb) *Site IIIS–20.*

FIGURE 14.—*Photographic Indexes to mission III near-side sites.*—Continued.

(cc) *Site IIIS-21.*

FIGURE 14.—*Photographic Indexes to mission III near-side sites.*—Continued.

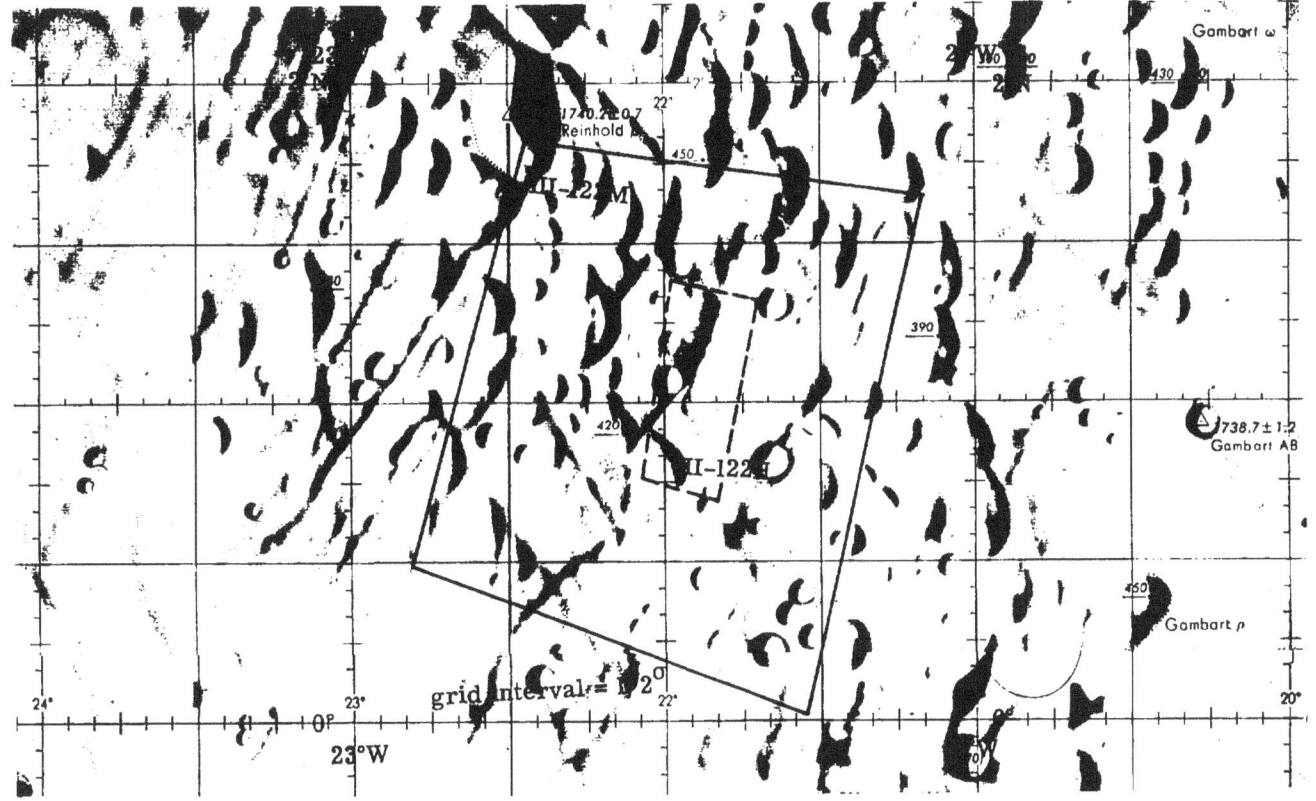

(dd) Site IIIS-22.

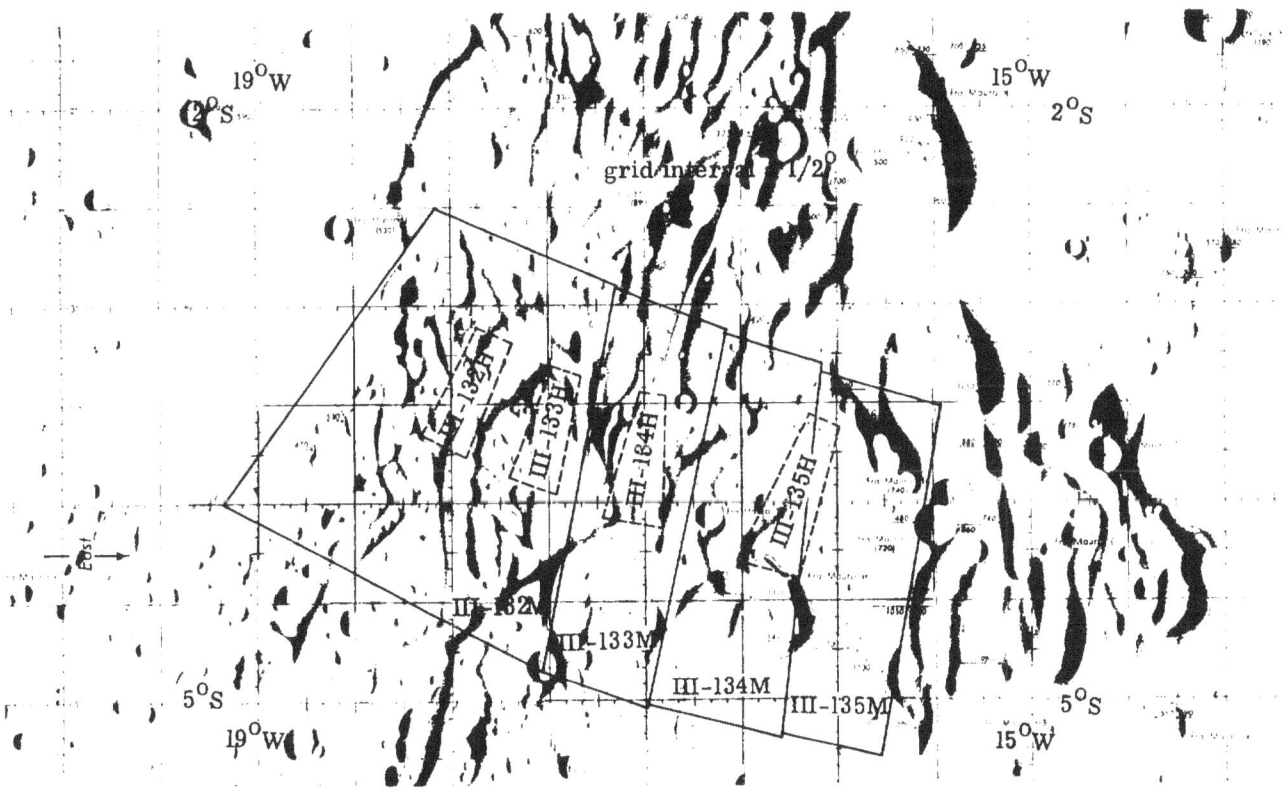

(ee) Site IIIS-23.

FIGURE 14.—Photographic Indexes to mission III near-side sites.—Continued.

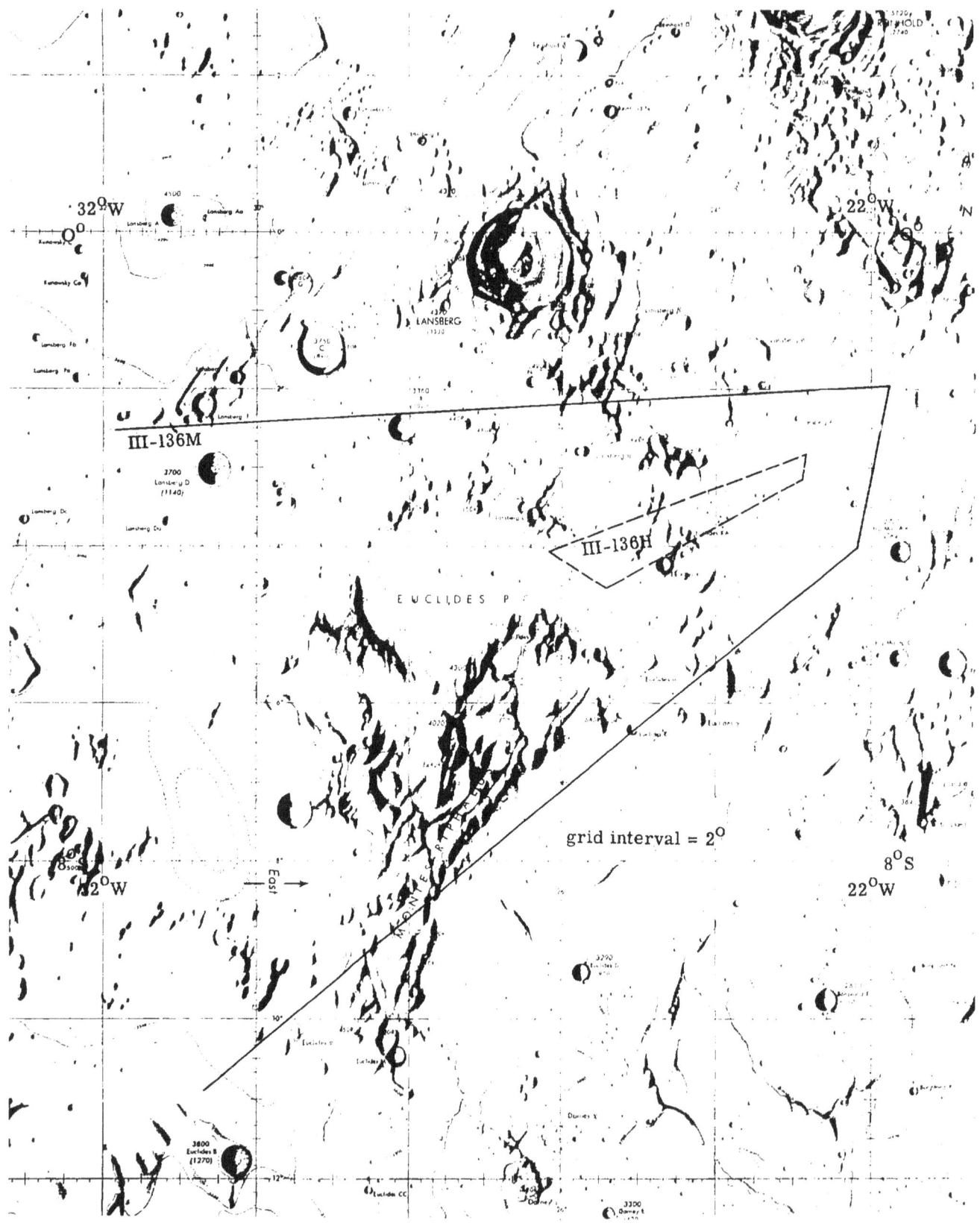

(ff) Site IIIS-24.

FIGURE 14.—*Photographic Indexes to mission III near-side sites.*—Continued.

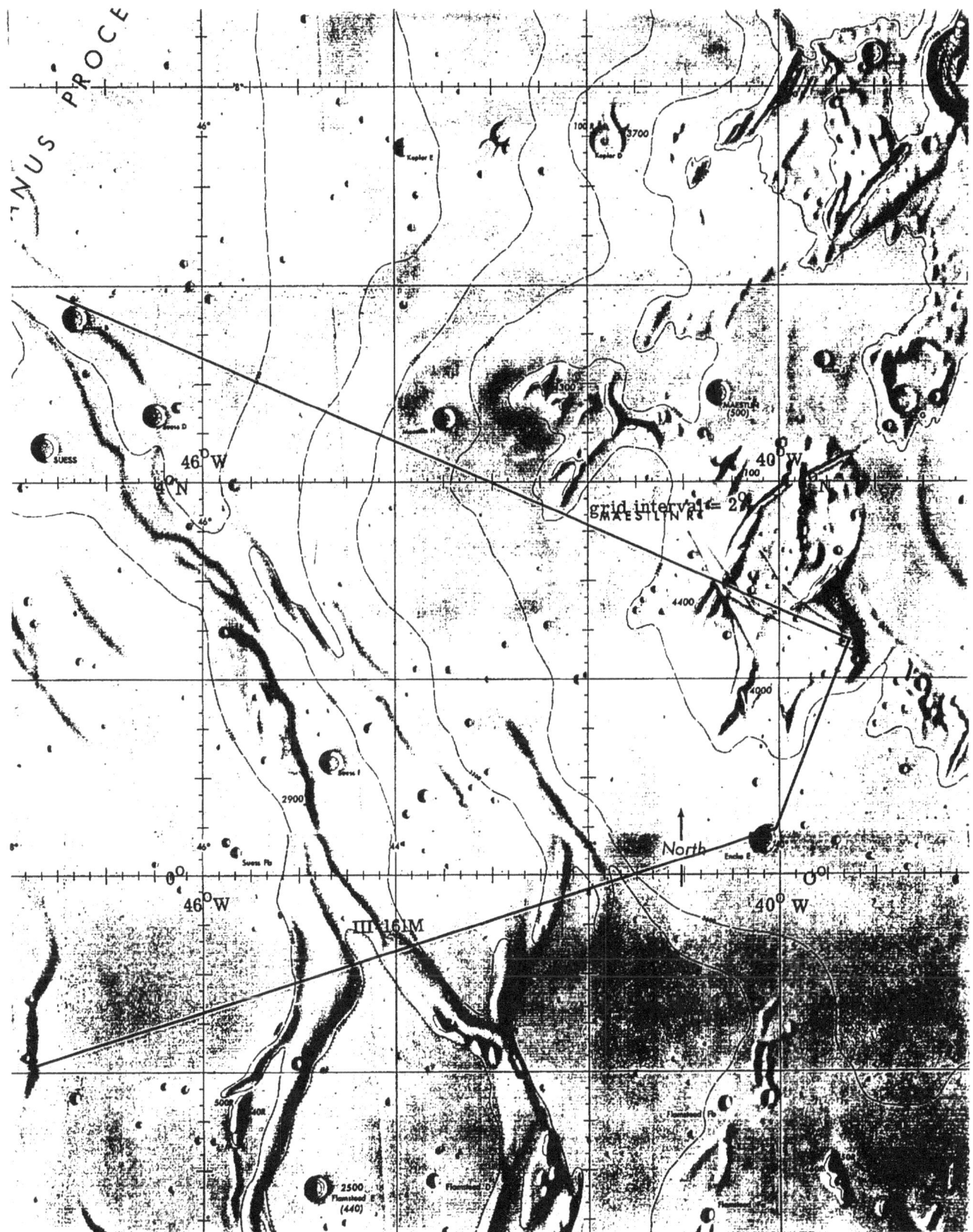

(gg) *Site IIIS–25.*

FIGURE 14.—*Photographic Indexes to mission III near-side sites.*—Continued.

(hh) *Site IIIS-26.*

FIGURE 14.—*Photographic Indexes to mission III near-side sites.*—Continued.

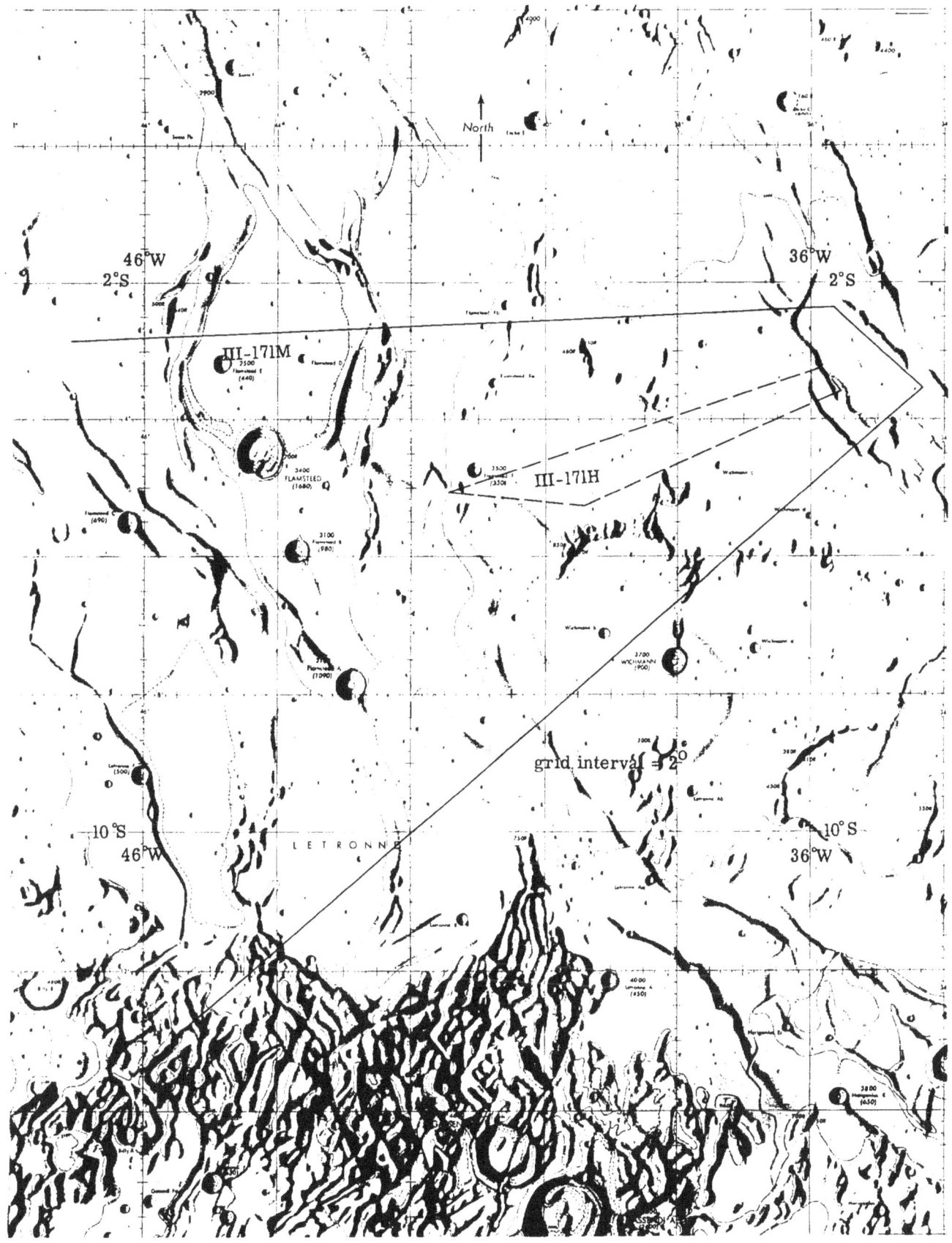

(ii) *Site IIIS-27.*

FIGURE 14.—*Photographic Indexes to mission III near-side sites.*—Continued.

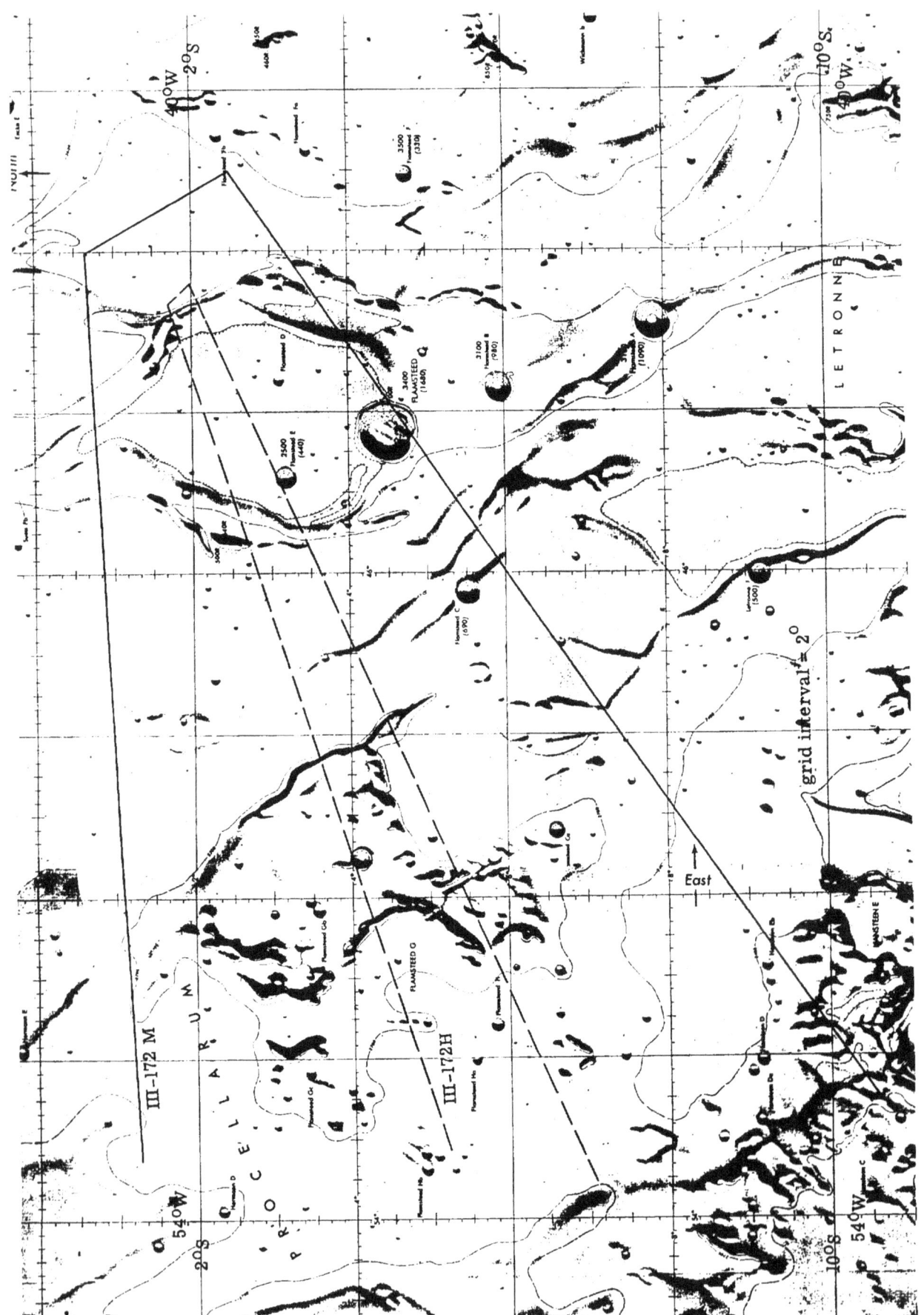

(j) *Site IIIS-28.*

FIGURE 14.—*Photographic Indexes to mission III near-side sites.*—Continued.

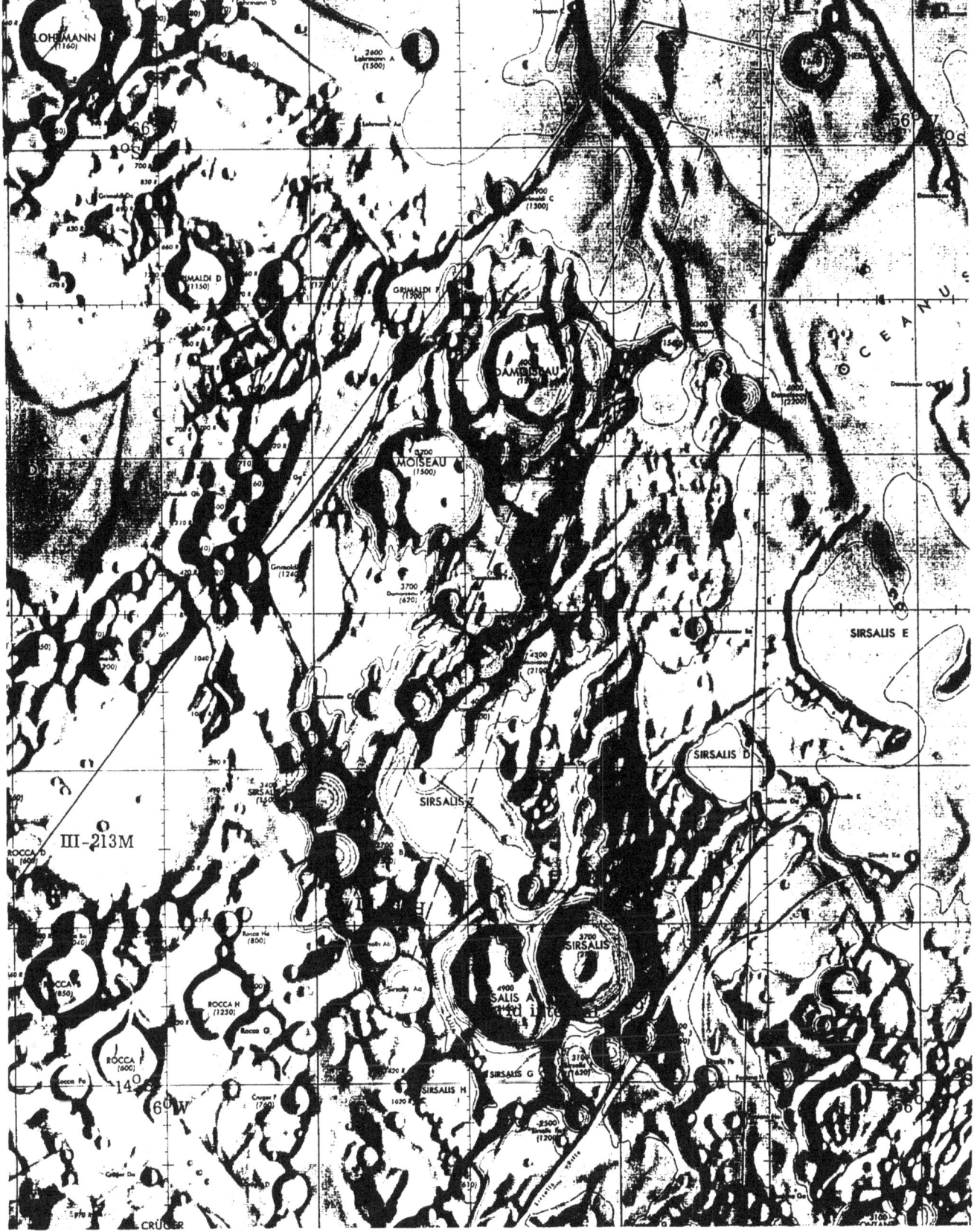

(kk) *Site IIIS-29.*

FIGURE 14.—*Photographic Indexes to mission III near-side sites.*—Continued.

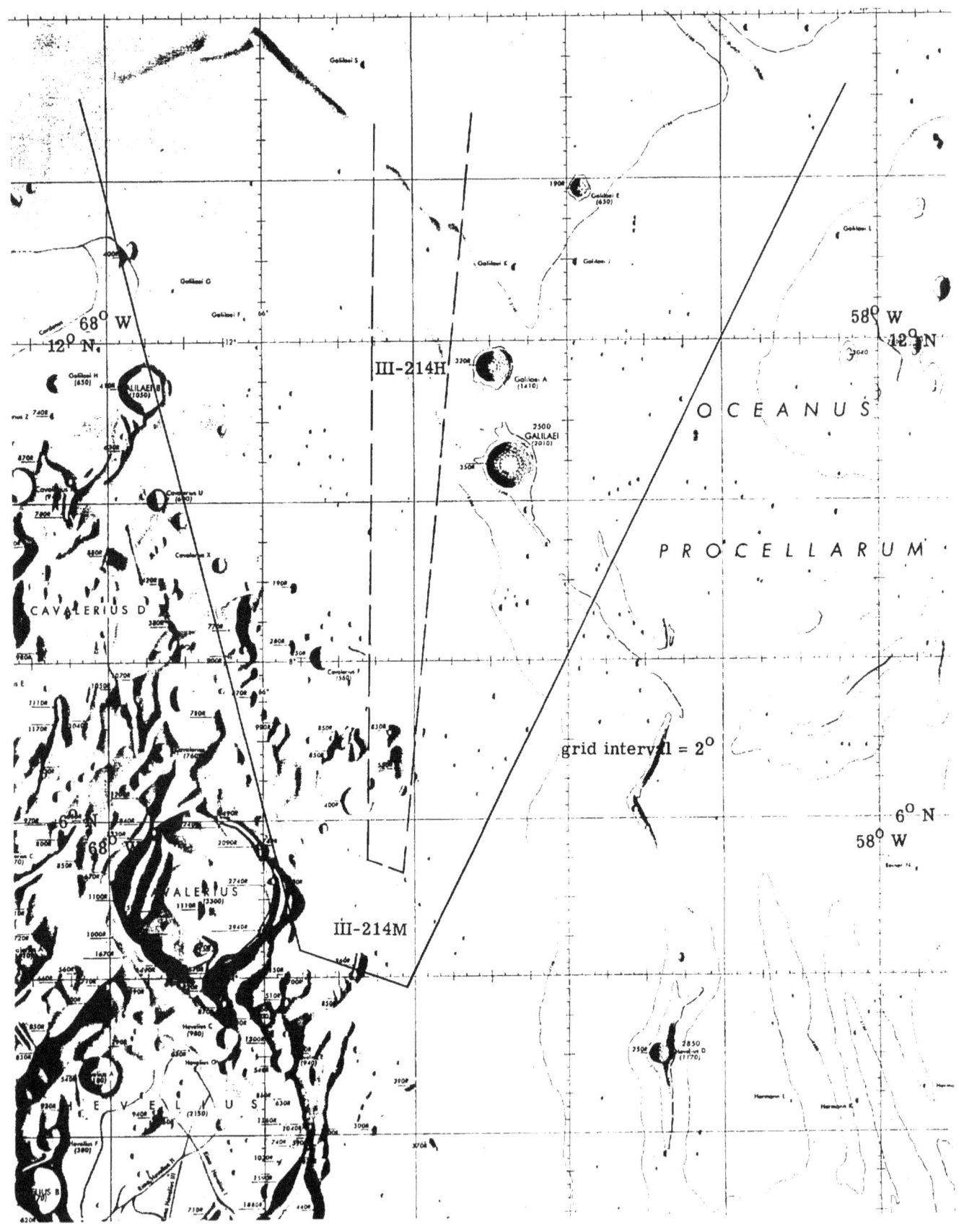

(11) *Site IIIS–30.*

FIGURE 14.—*Photographic Indexes to mission III near-side sites.*—Continued.

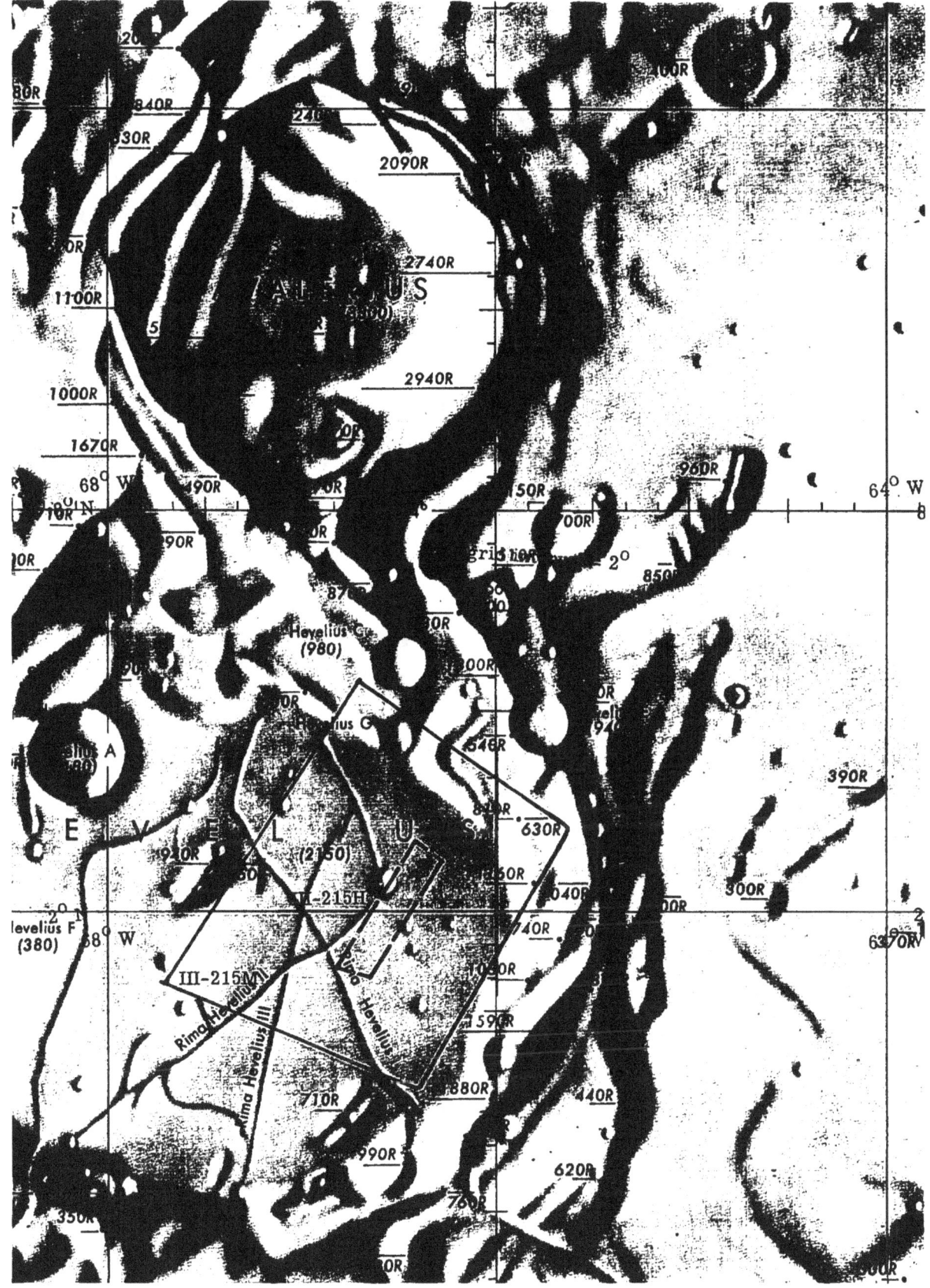

(mm) *Site IIIS-31.*

FIGURE 14.—*Photographic Indexes to mission III near-side sites.*—Concluded.

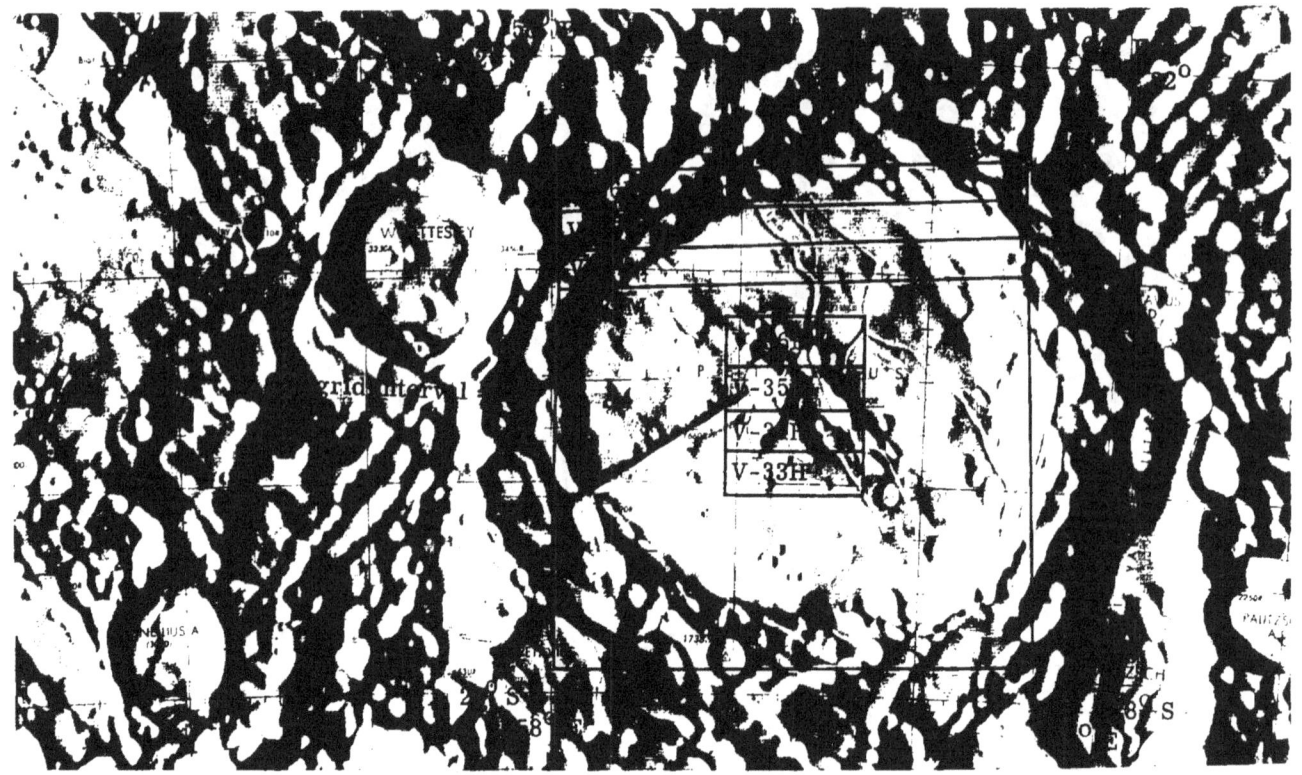

(a) *Site V–1.*

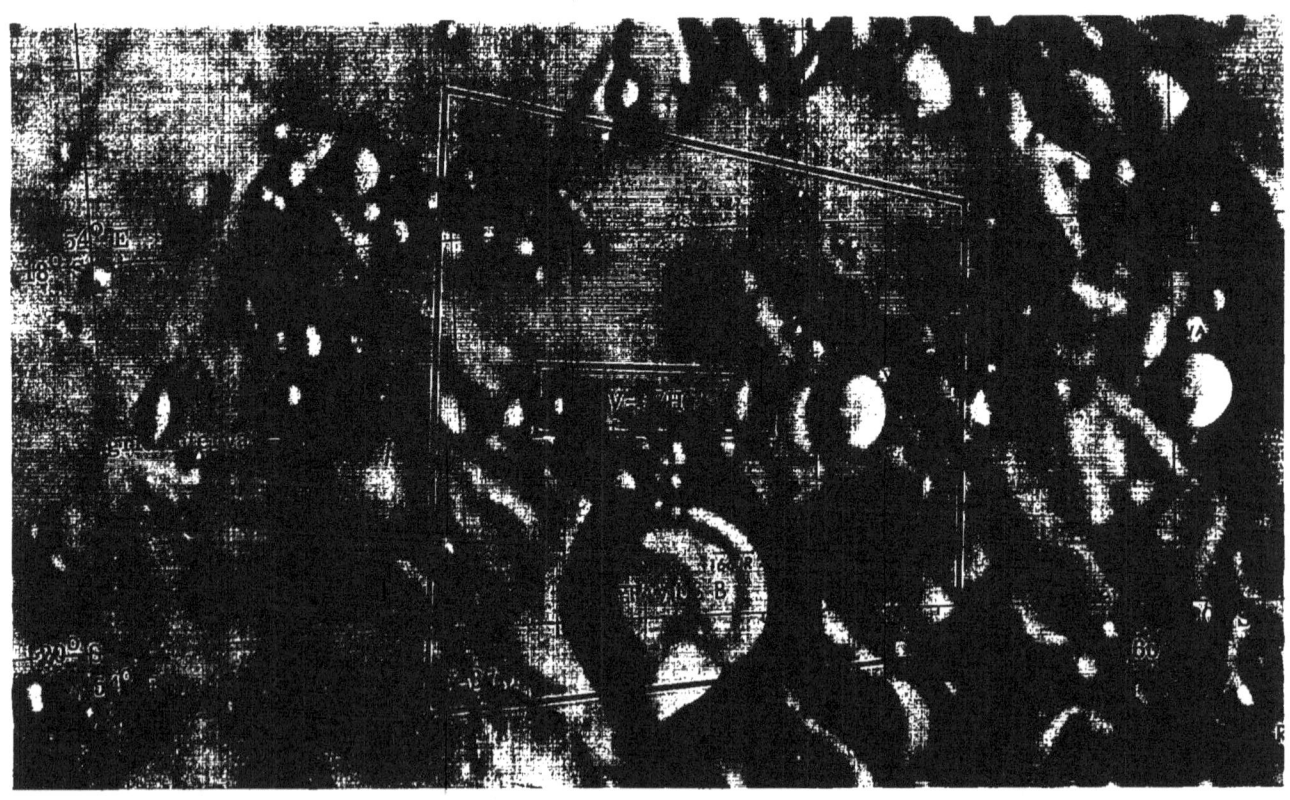

(b) *Site V–2.1.*

FIGURE 15.—*Photographic Indexes to mission V near-side sites.*

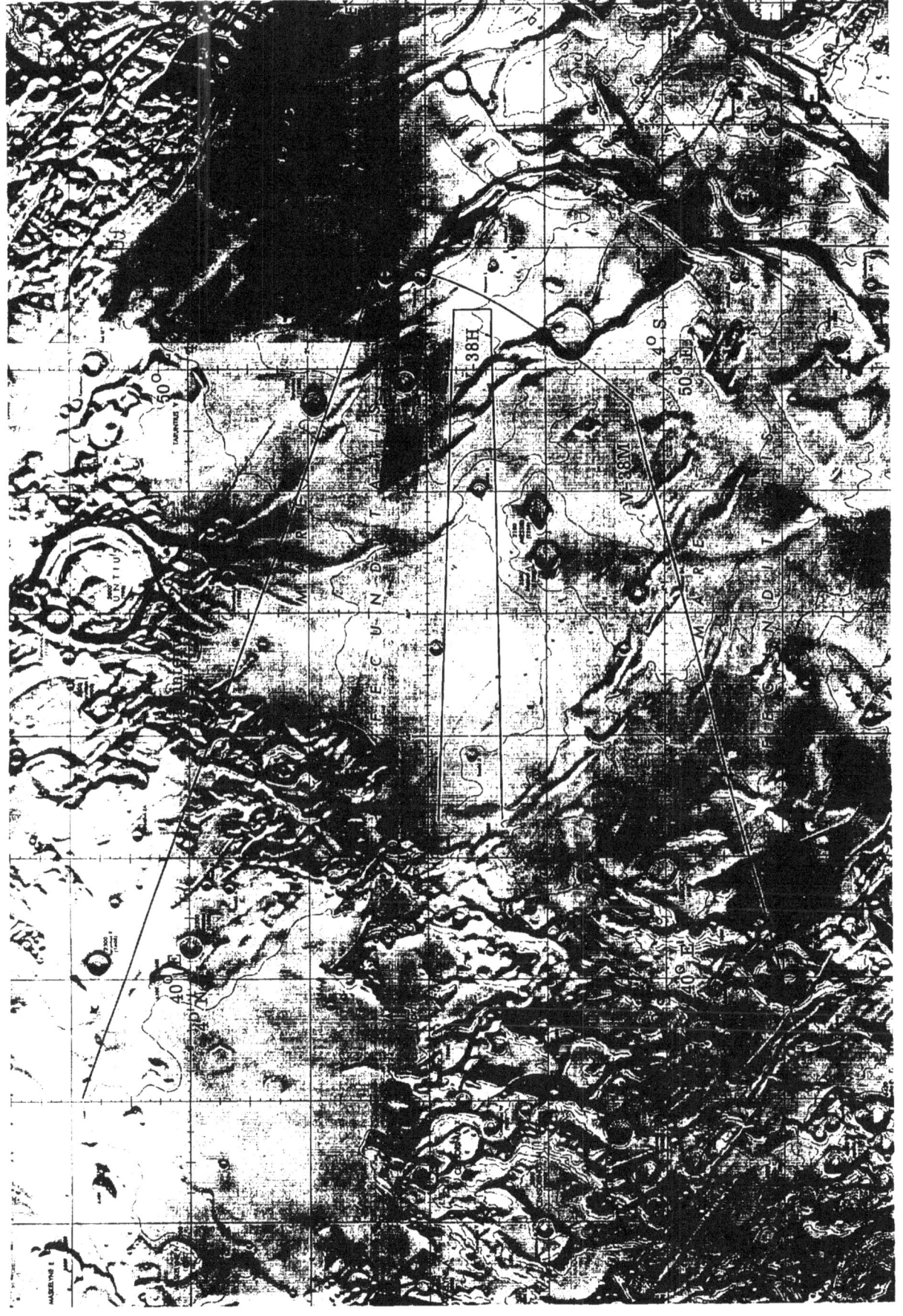

(c) *Site I'–3.1.*

FIGURE 15.—*Photographic Indexes to mission I' near-side sites.*—Continued.

(e) Site V-8.

(d) Site V-4.

FIGURE 15.—Photographic Indexes to mission V near-side sites.—Continued.

102

M A R E

grid interval = 2°

50° E   2° N

4° S

50° E

F
E
C
U
N
D
I
T
A
T
I
S

M A R E

F
E
C
U
N
D
I
T
A
T
I
S

V'-41H

V-41M

North

38° E

2° N

4° S

38° E

(f) *Site V-51.*

FIGURE 15.—*Photographic Indexes to mission V near-side sites.*—Continued.

103

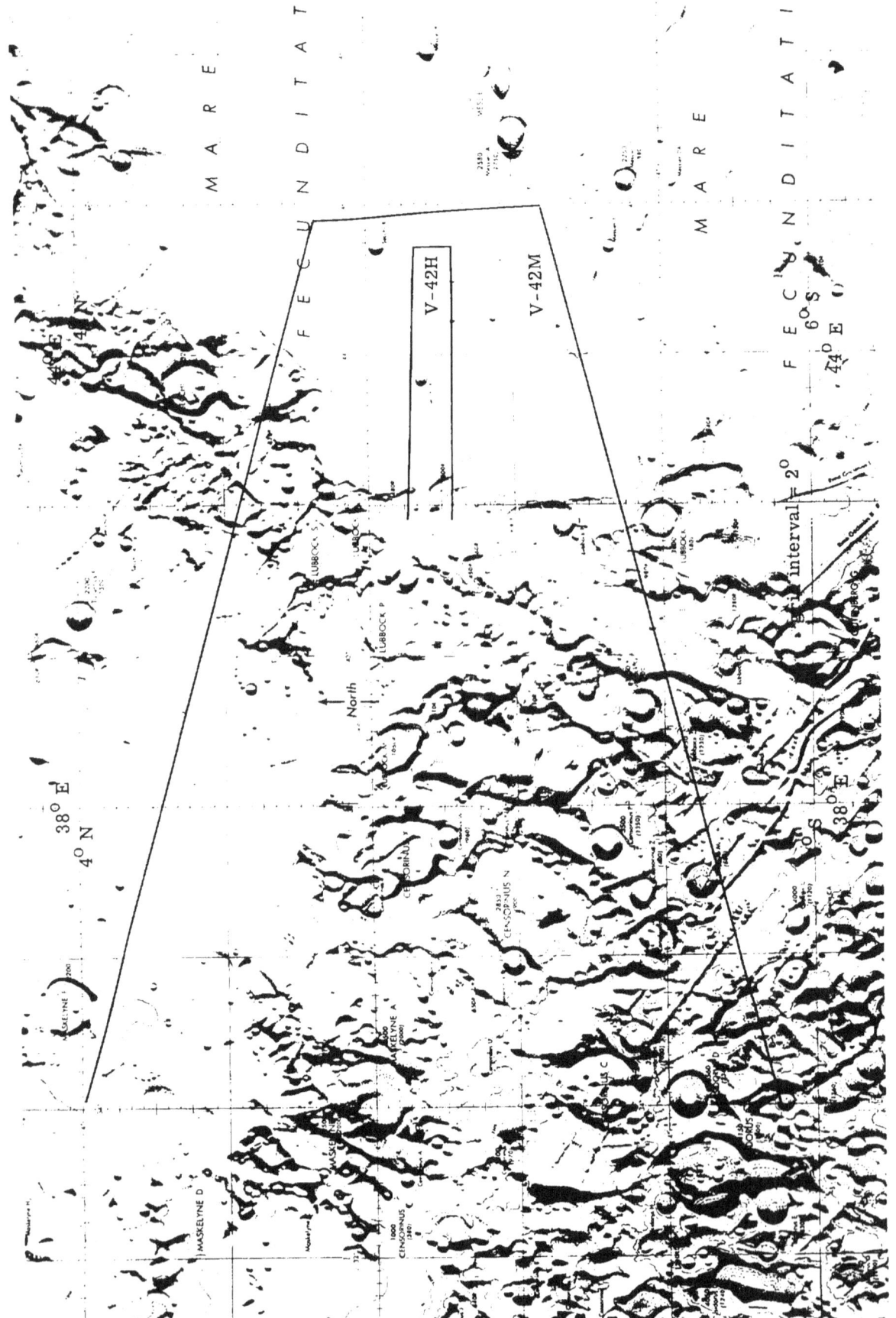

(g) *Site V–6.*

FIGURE 15.—*Photographic Indexes to mission V near-side sites.*—Continued.

(h) *Site V-9.1.*

FIGURE 15.—*Photographic Indexes to mission V near-side sites.*—Continued.

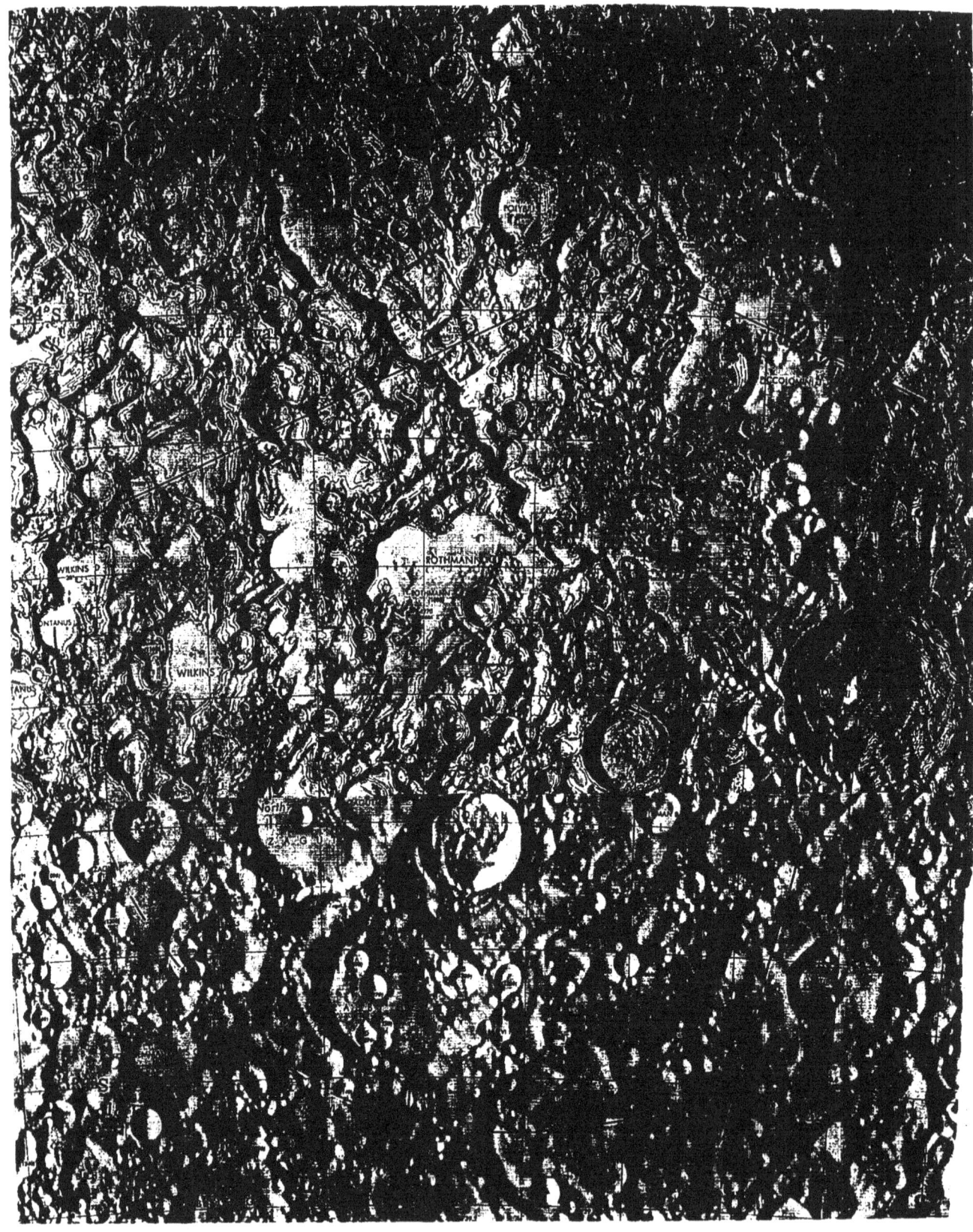

(i) *Site V-10.*

FIGURE 15.—*Photographic Indexes to mission V near-side sites.*—Continued.

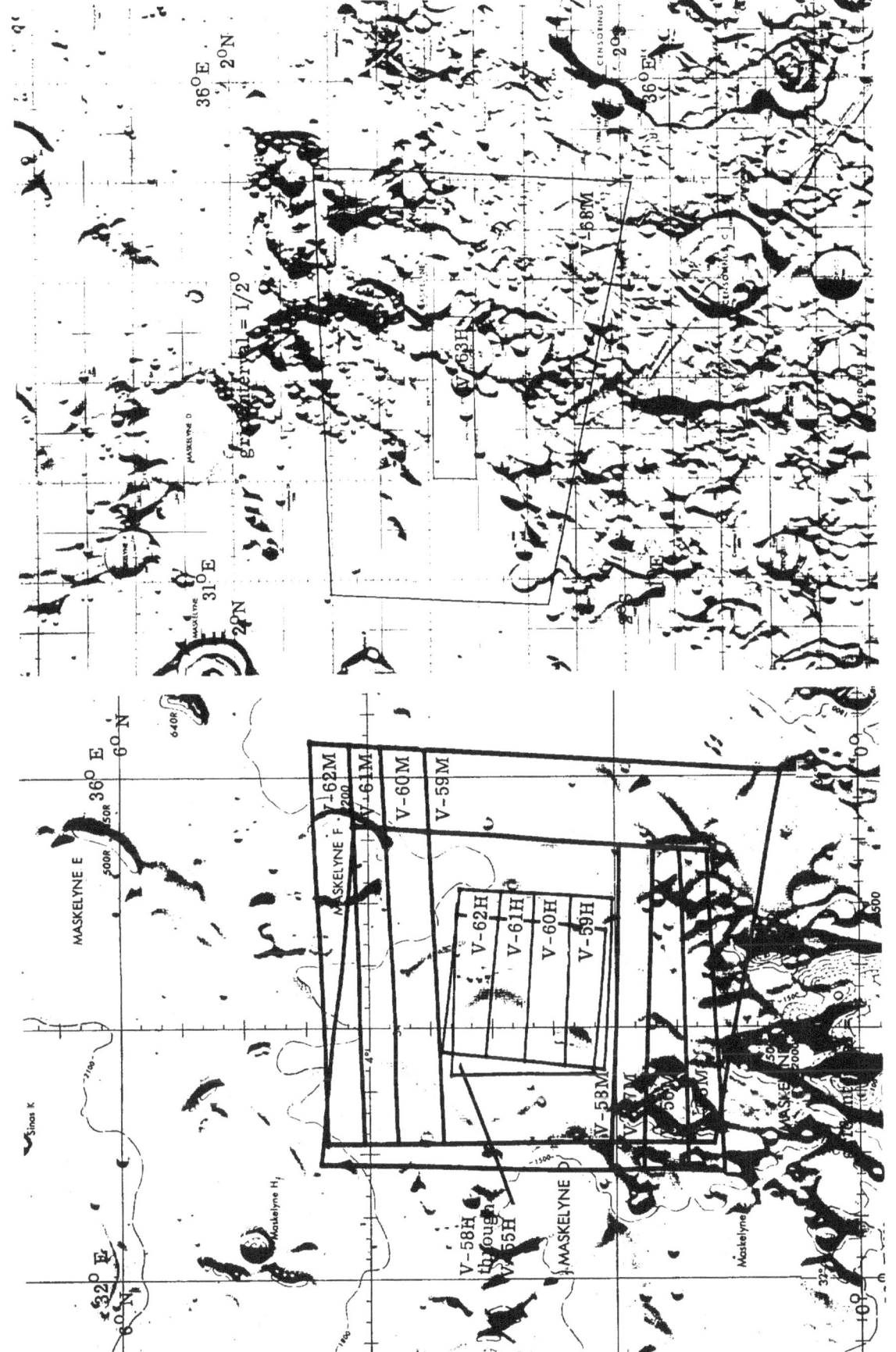

(k) *Site V–12.*

(j) *Site V–11.*

Figure 15.—*Photographic Indexes to mission V near-side sites.—Continued.*

107

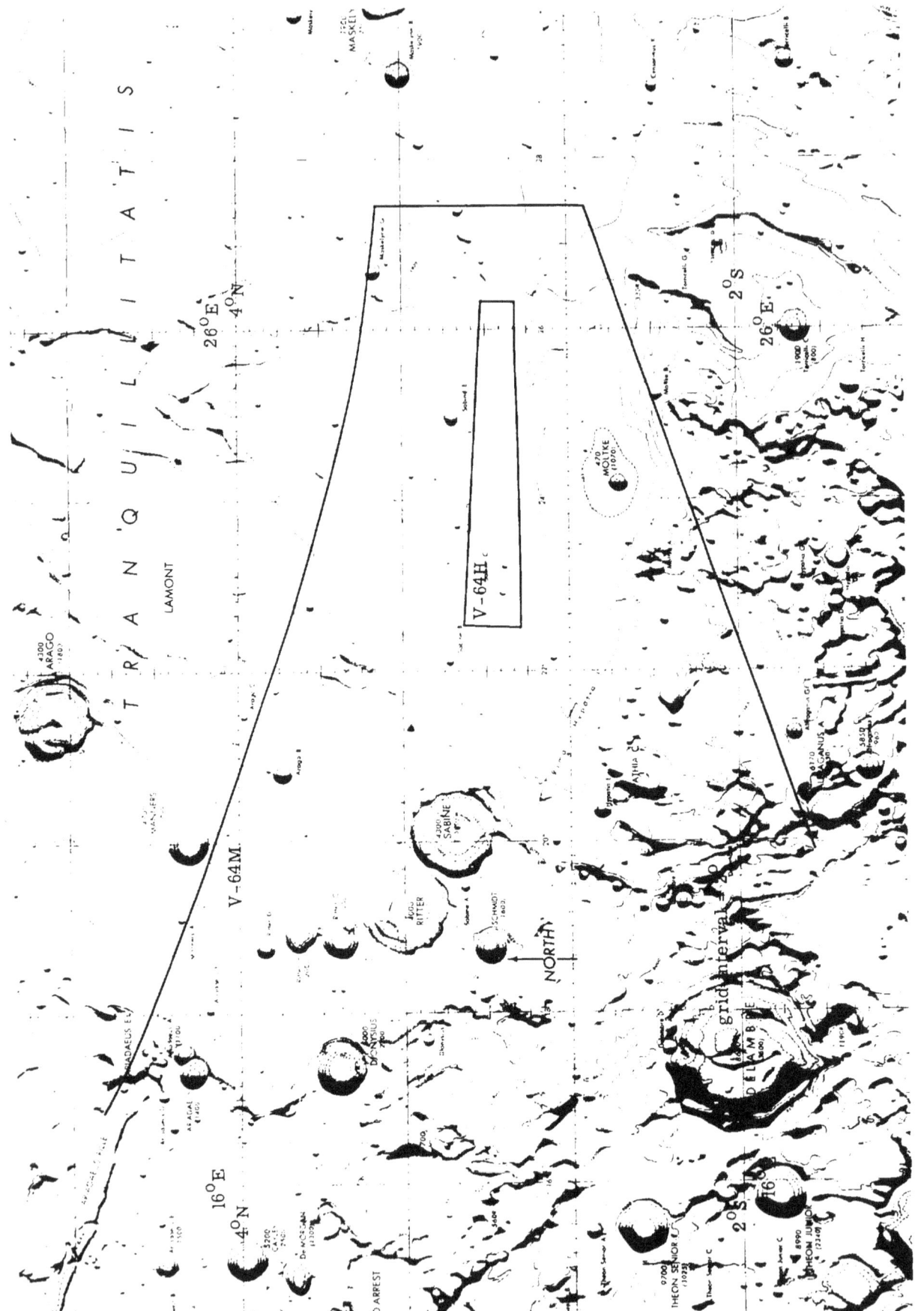

Figure 15.—Photographic indexes to mission V near-side sites.—Continued.

(1) Site V-13.

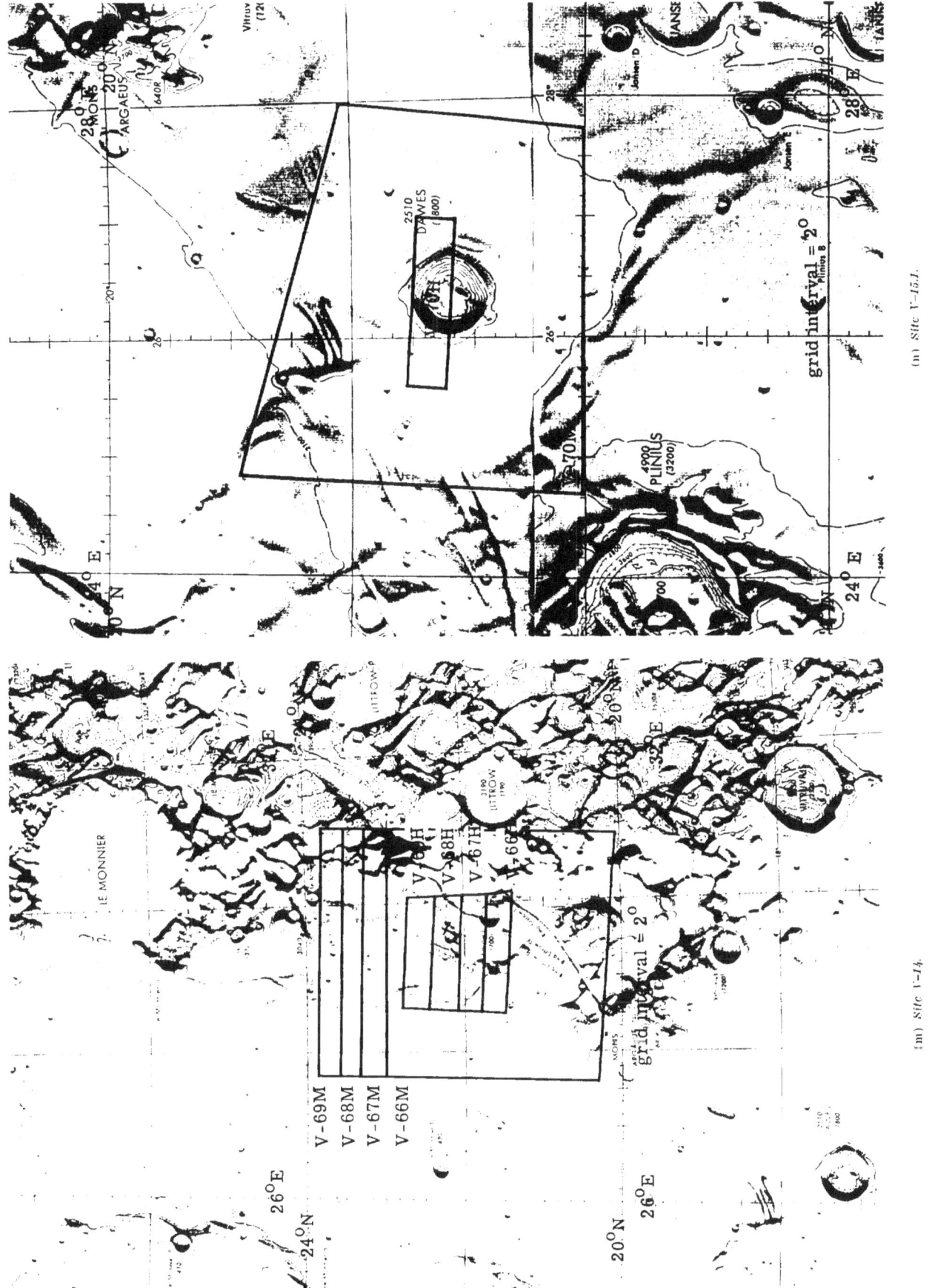

(n) Site V-15.1.

(m) Site V-14.

FIGURE 15.—Photographic Indexes to mission V near-side sites.—Continued.

(b) Site V-18.

(a) Site V-16.

FIGURE 15.—Photographic indexes to mission V near-side sites.—Continued.

(r) Site V–21.

(q) Site I–19.

FIGURE 15.—*Photographic indexes to mission V near-side sites.—Continued.*

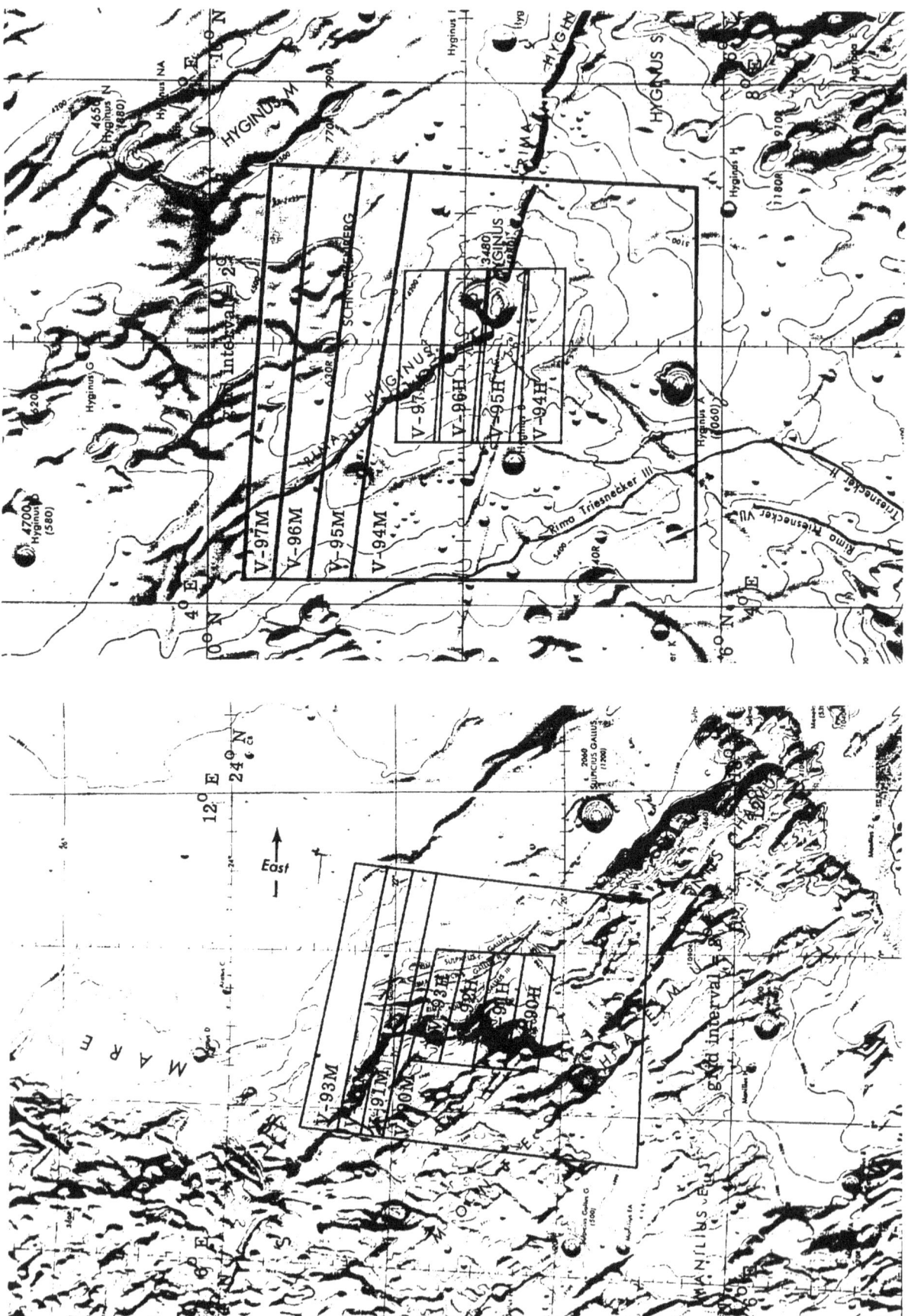

(t) *Site V–23.1.*

(s) *Site V–22.*

FIGURE 15.—*Photographic Indexes to mission V near-side sites.—Continued.*

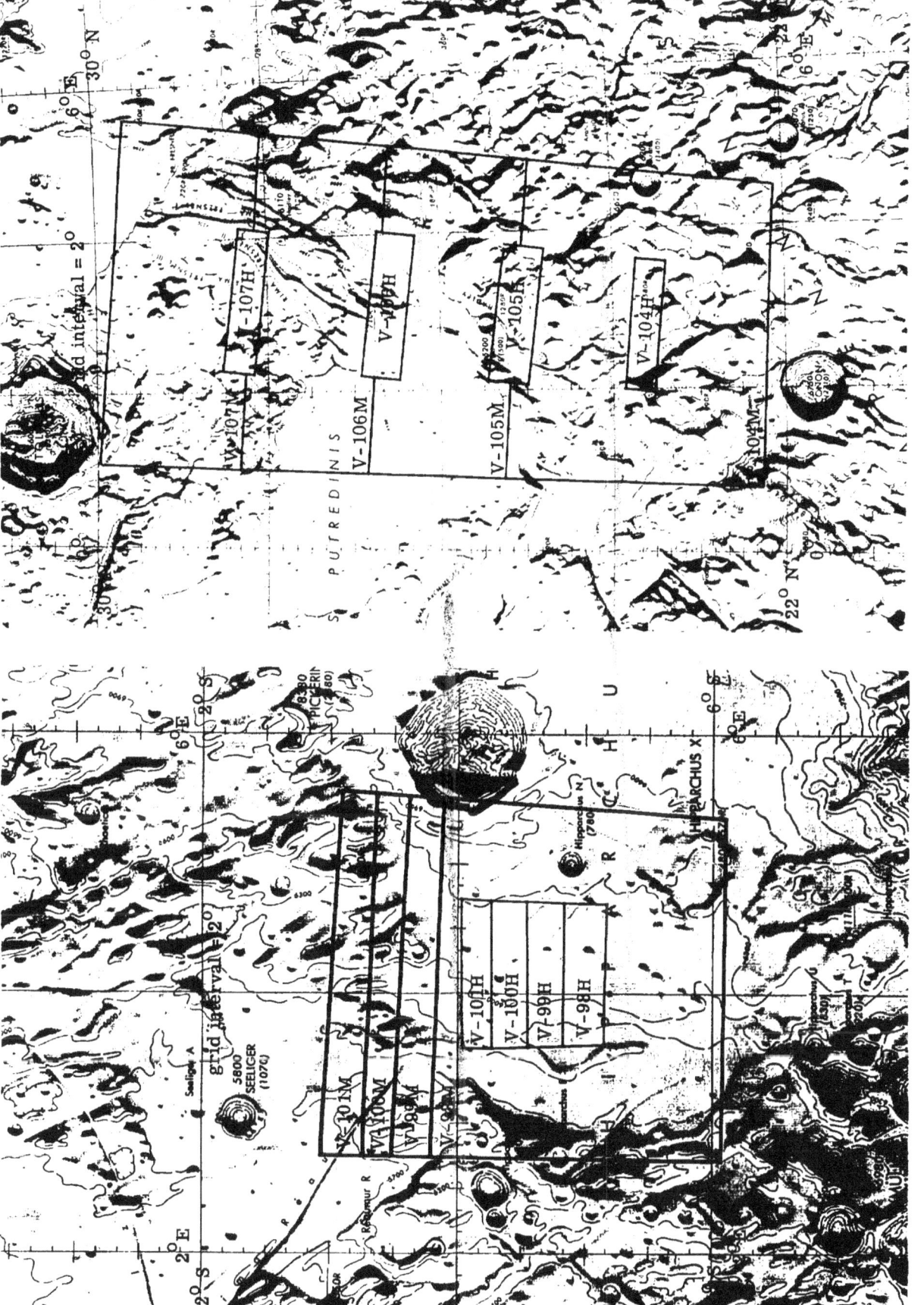

(v) *Site V-26.1.*

(u) *Site V-24.*

FIGURE 15.—*Photographic indexes to mission V near-side sites.*—Continued.

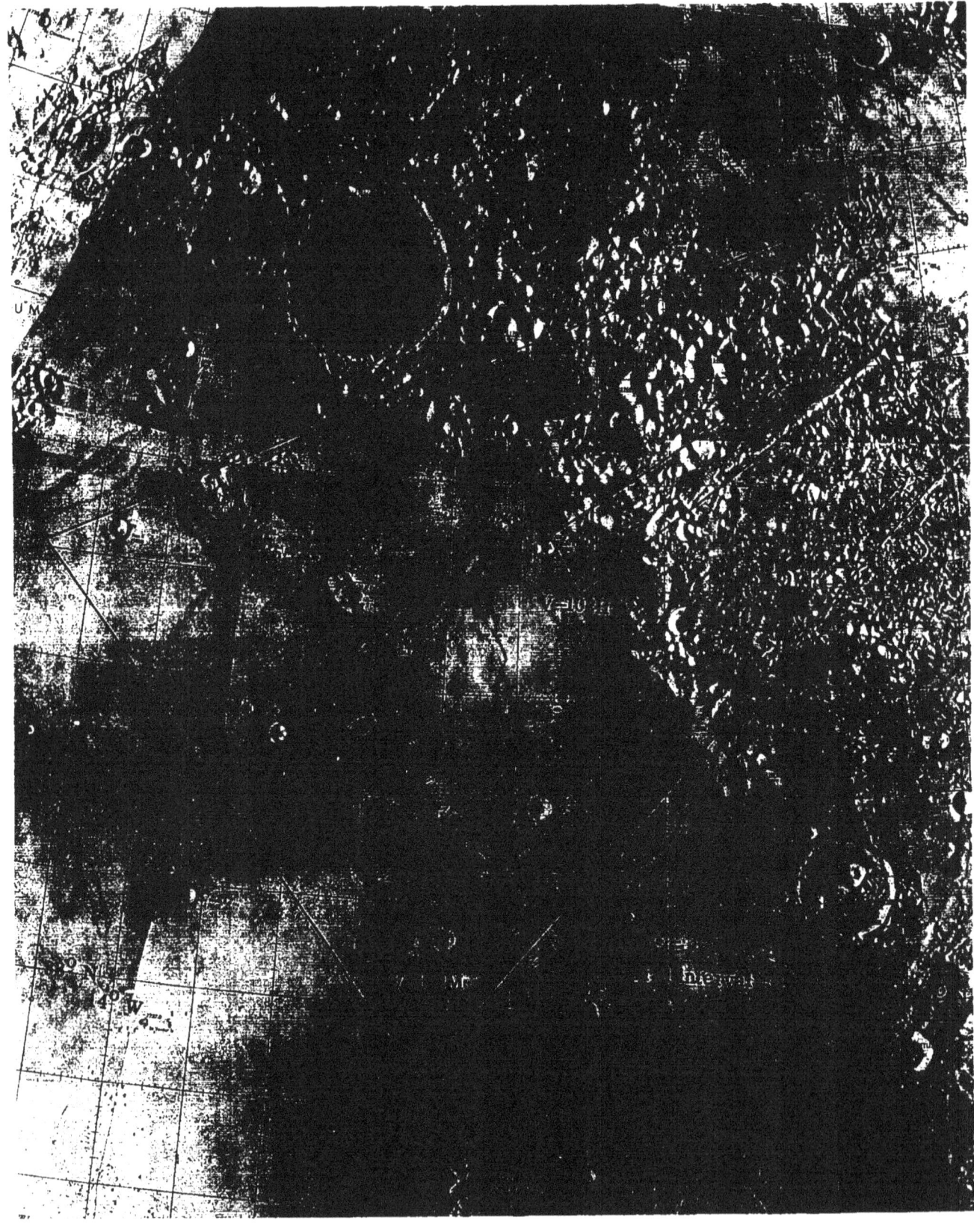

(w) *Site V–25.*

FIGURE 15.—*Photographic Indexes to mission V near-side sites.*—Continued.

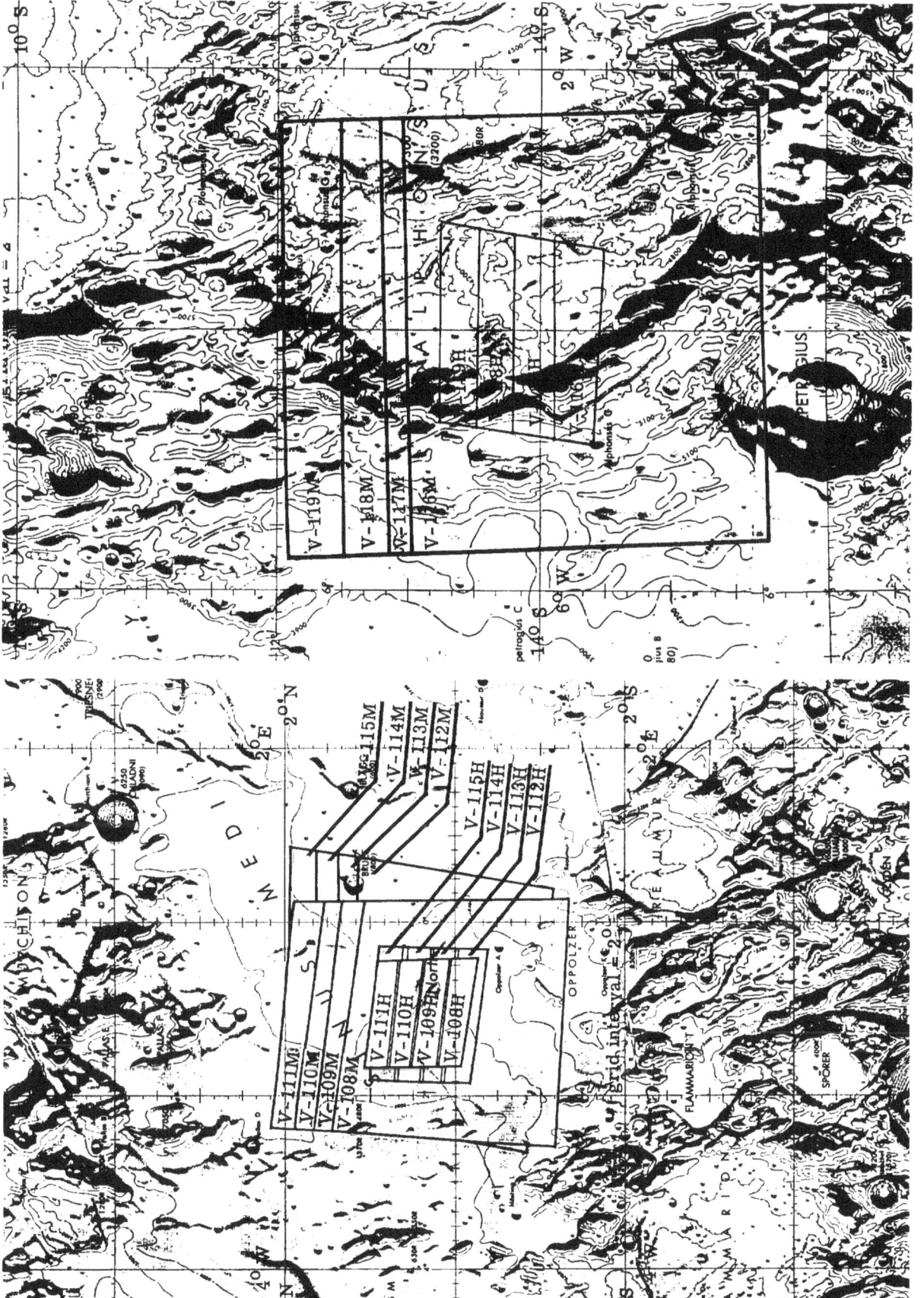

(y) Site V-28.

(x) Site V-27.

FIGURE 15.—*Photographic indexes to mission V near-side sites.*—Continued.

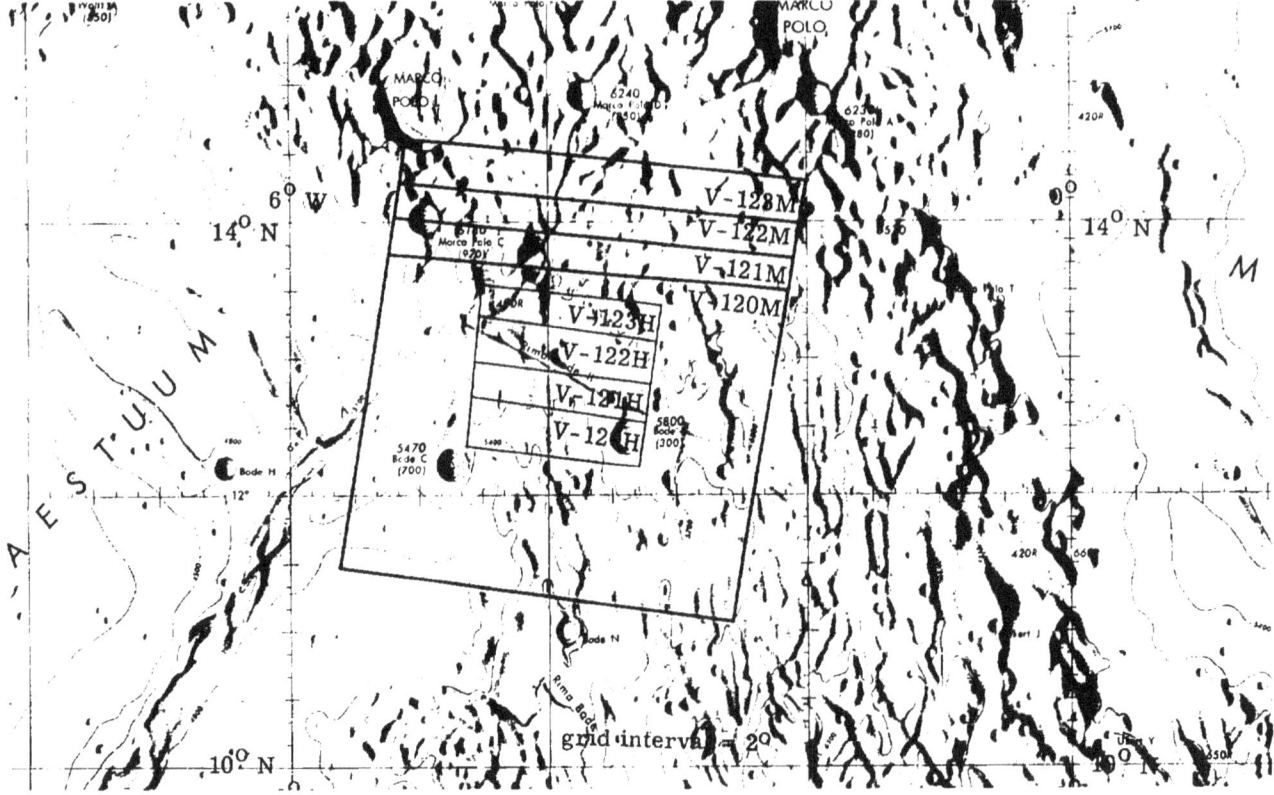

(z) *Site V-29.*

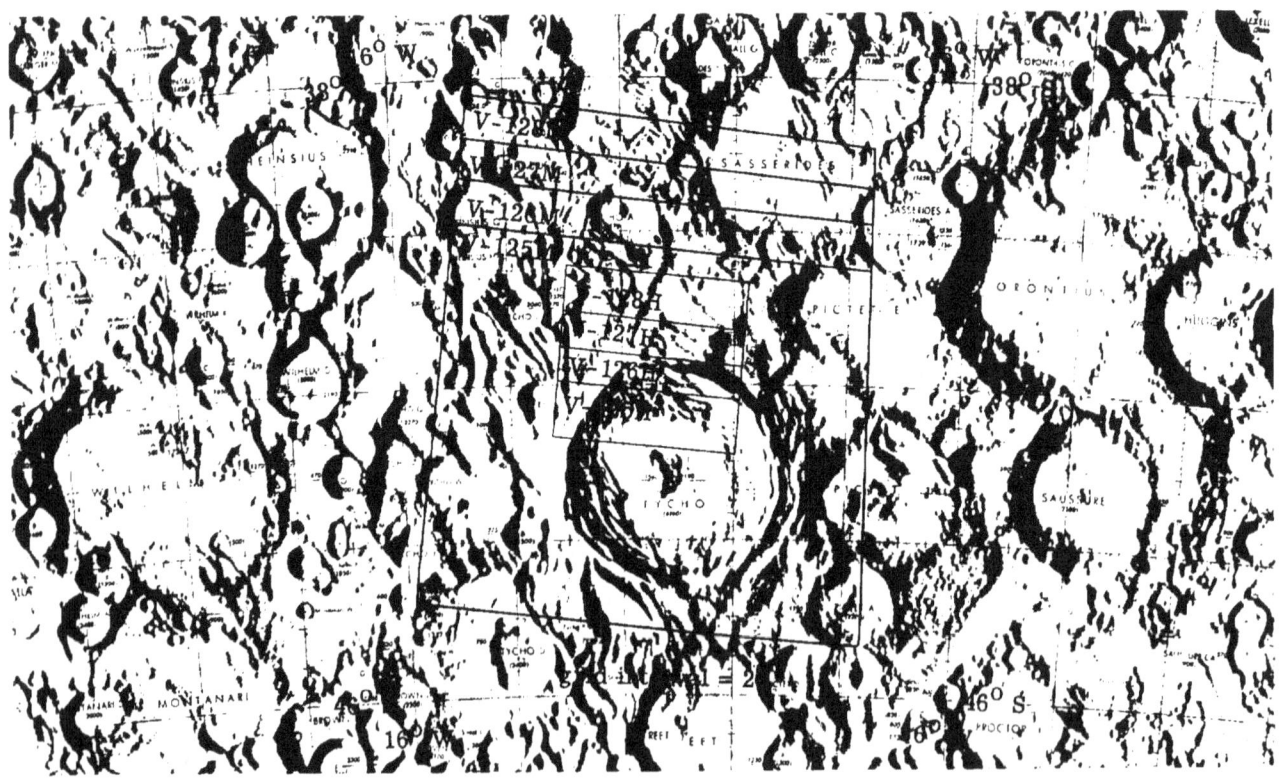

(aa) *Site V-30.*

FIGURE 15.—*Photographic Indexes to mission V near-side sites.*—Continued.

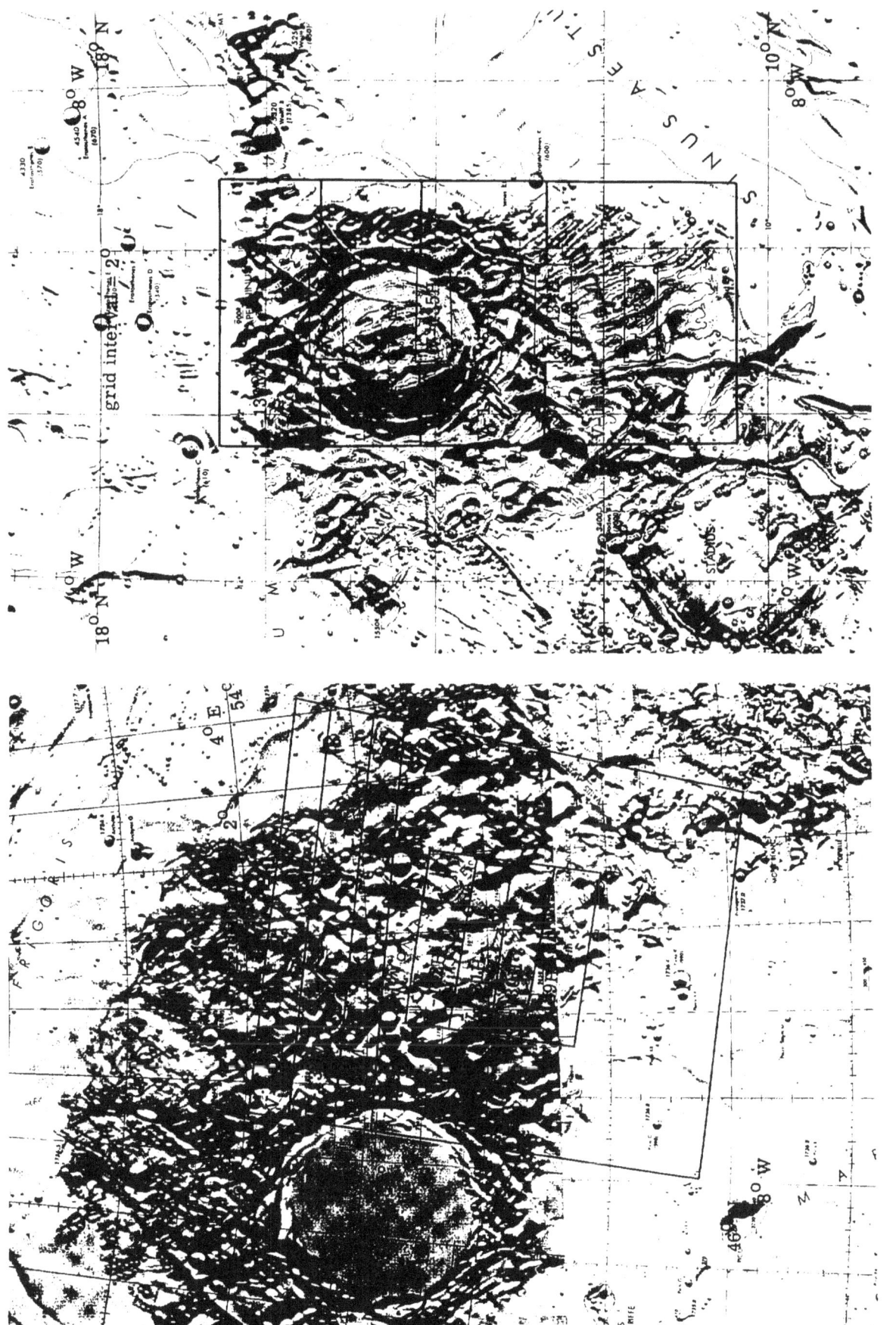

(cc) *Site Γ-32.*

(bb) *Site Γ-31.*

FIGURE 15.—*Photographic Indexes to mission Γ near-side sites.*—Continued.

(ee) *Site V-34.*

(dd) *Site V-33.*

FIGURE 15.—*Photographic Indexes to mission V near-side sites.*—Continued.

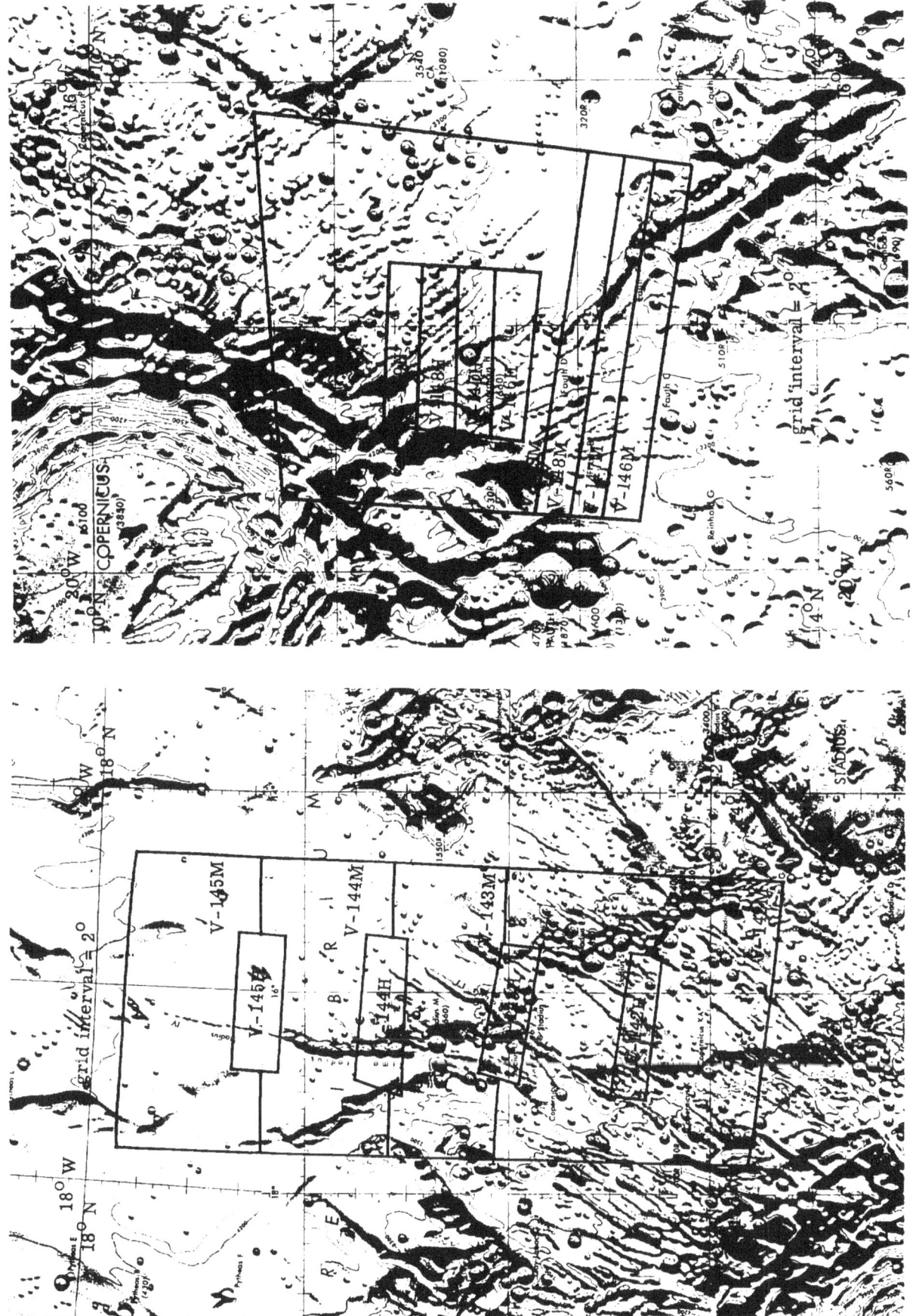

(gg) Site V-96.

(ff) Site V-35.

FIGURE 15.—*Photographic Indexes to mission V near-side sites.*—Continued.

119

grid interval = 2°

18° W
36° N

V-162M
V-161M
V-160M
V-159M

North

V-162H
V-161H
V-160H
V-159H

R E
26° W
36° N

30° N
26° W

30° N

30° N
18° W

(ii) *Site V-38.*

COPERNICUS

(hh) *Site V-37.*

FIGURE 15.—*Photographic Indexes to mission V near-side sites.—Continued.*

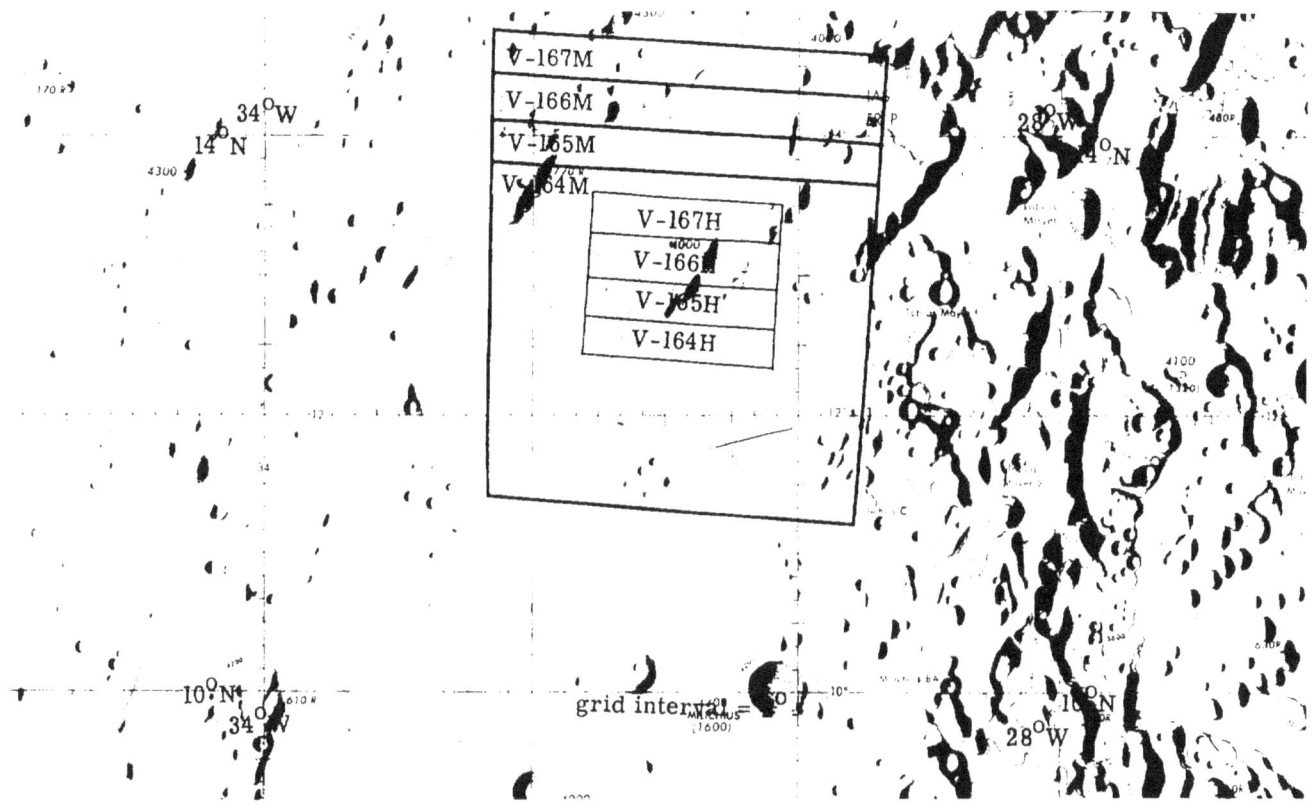

(jj) *Site V-40.*

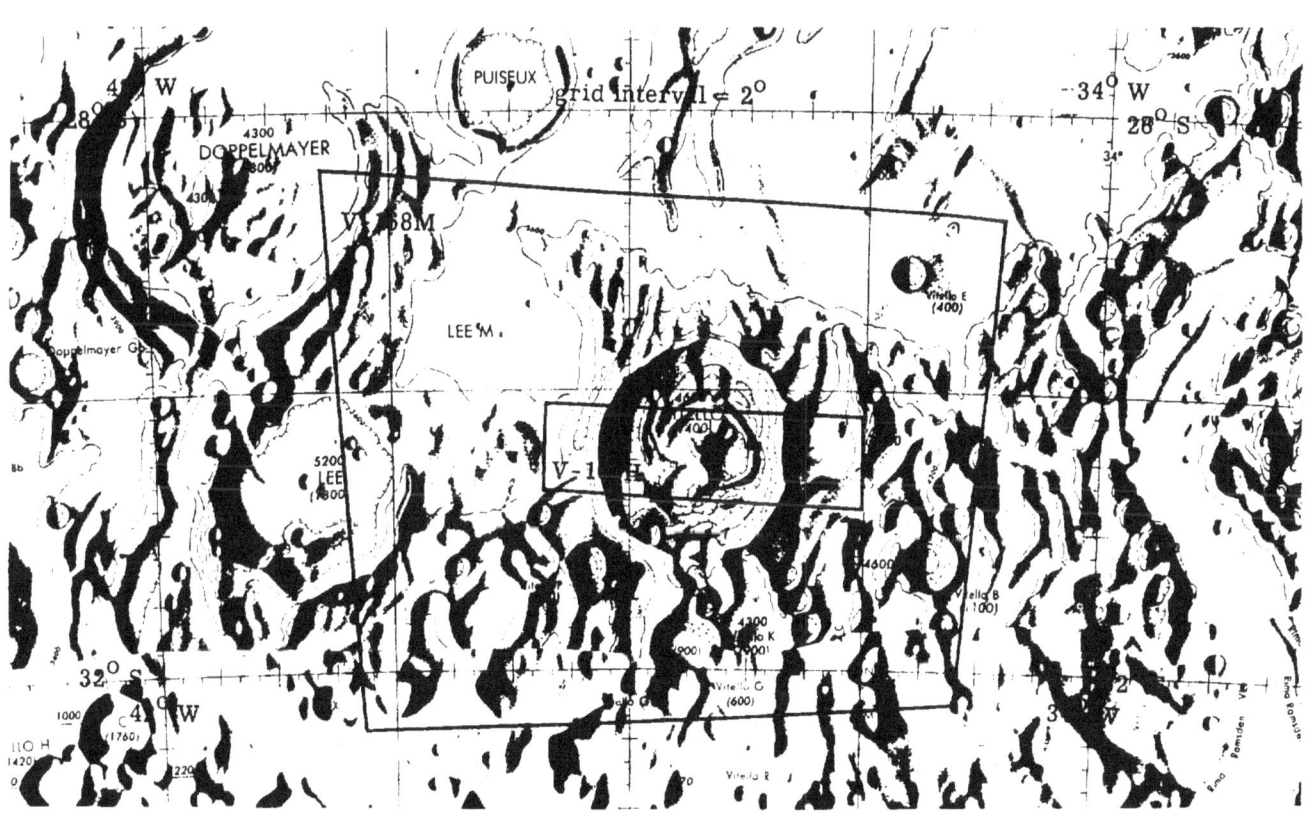

(kk) *Site V-41.*

FIGURE 15.—*Photographic Indexes to mission V near-side sites.*—Continued.

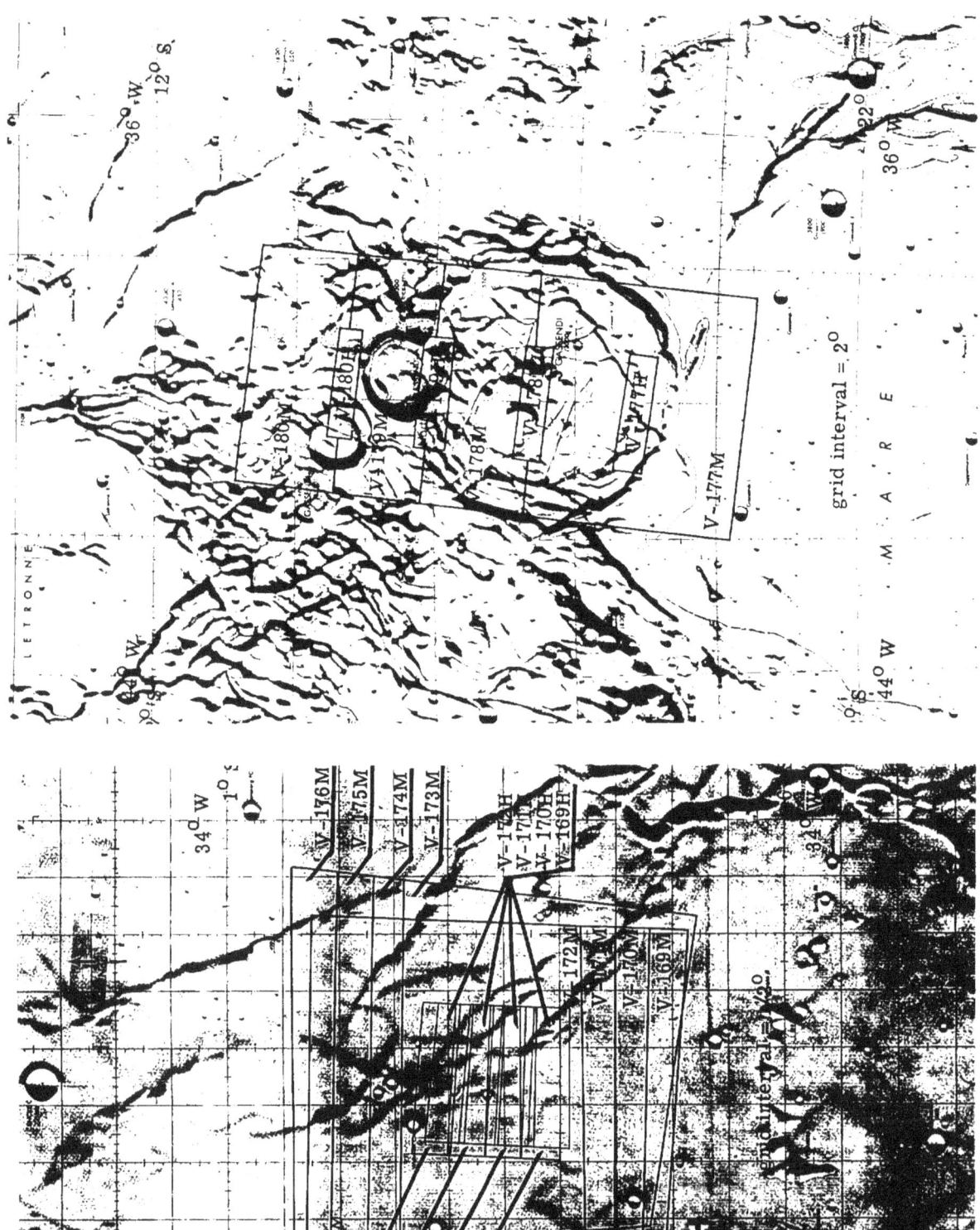

(mm) *Site V-43.2.*

grid interval = 2°

(11) *Site V-42.*

FIGURE 15.—*Photographic indexes to mission V near-side sites.*—Continued.

(oo) Site V-46.

(nn) Site V-45.1.

FIGURE 15.—Photographic Indexes to mission V near-side sites.—Continued.

(b) Site V-49.

(a) Site V-48.

FIGURE 15.—Photographic indexes to mission V near-side sites.—Continued.

(ss) *Site V–51.*

(rr) *Site V–50.*

Figure 15.—*Photographic Indexes to mission V near-side sites.*—Concluded.

125

www.ingramcontent.com/pod-product-compliance
Lightning Source LLC
Chambersburg PA
CBHW080259180526

45167CB00006B/2593